RECENT ADVANCES IN PLASMA DIAGNOSTICS

DIAGNOSTIKA PLASMY

ДИАГНОСТИКА ПЛАЗМЫ

RECENT ADVANCES IN PLASMA DIAGNOSTICS

Volume 3

Corpuscular, Correlation, Bolometric, and Other Techniques

Edited by V. T. Tolok

Engineering Physics Institute of the Academy of Sciences of the
Ukrainian SSR, Khar'kov, and Chairman, Department of Plasma
Physics, Khar'kov State University

Translated from Russian by Joachim R. Büchner

CONSULTANTS BUREAU · NEW YORK-LONDON · 1971

This volume comprises the translation of pages 338-443 of *Diagnostika Plasmy, Vyp. 2*, originally published by Atomizdat in Moscow in 1968. The translation of pages 1-187 was published in Volume 1 and that of pages 188-337 in Volume 2 of this series.

The original Russian text has been corrected by the editor, and the translation is published under an agreement with Mezhdunarodnaya Kniga, the Soviet book export agency.

B. T. Tolok

ДИАГНОСТИКА ПЛАЗМЫ

Library of Congress Catalog Card Number 70-140828

ISBN 978-1-4684-7579-1   ISBN 978-1-4684-7577-7 (eBook)
DOI 10.1007/978-1-4684-7577-7

© 1971 Consultants Bureau, New York
Softcover reprint of the hardcover 1st edition 1971
A Division of Plenum Publishing Corporation
227 West 17th Street, New York, N.Y. 10011

United Kingdom edition published by Consultants Bureau, London
A Division of Plenum Publishing Company, Ltd.
Davis House (4th Floor), 8 Scrubs Lane, Harlesden, NW10 6SE, England

CONTENTS

MEASUREMENTS OF THE ION TEMPERATURE
IN PLASMAS FROM NEUTRON RADIATION

S. P. Bogdanov and V. I. Volosov

Methods employing the neutron radiation generated in nuclear reactions in the plasma are of great importance for the diagnosis of the ion component of a hot *dd* or *dt* plasma. These methods permit a rather accurate determination of the energy state of the ion component, eliminate the possibilities for certain errors, and are characterized by the fact that the recording circuits of the apparatus are situated outside the vacuum chamber.

Measurements of the ion temperature from the integral neutron output from the plasma are well established at the present time [1, 2]. When the plasma density is determined with some other independent method, one can compute the ion temperature from a comparison of the results of the two experiments. However, the integral method is appropriate only for a plasma with a Maxwell velocity distribution of the ions or for rather hot (>10 keV) plasmas. This condition results from the strong dependence of the neutron output on the form of the ion spectrum at $T_i \leq 10$ keV (with the average ion energy assumed constant). For example, at an average ion energy $\bar{E} = 2$ keV, the integral ion output can change by a factor of 50 due to changes in the energy distribution function of the neutrons [3]. In addition, for calculating $\bar{E}$ one must know with sufficient accuracy the size of the region from which radiation is emitted. In the majority of experiments, it is extremely difficult to obtain this information.

Let us consider in detail another diagnostic method in which the form of the neutron spectrum (n spectrum) measured is used to determine the energy of the ion component. One must bear in mind that the width of the n spectrum is much greater than the width of the ion spectrum. This can be easily shown from the following estimate: assume that u denotes the average velocity of the plasma ions and v_n the velocity of a neutron in the center-of-mass system. Then the half-width of the neutron spectrum in the laboratory system has the following order of magnitude:

$$\Delta E_n = \frac{m}{2}[(v_n + u)^2 - v_n^2] = m v_n u$$

or

$$\Delta E_n = 2\sqrt{E_n k T_i}.$$

Modern methods of fast neutron spectrometry have an accuracy of 3-5%. Thus, it follows from the above estimate that the minimum plasma temperature which can be measured by this method is rather low (about 1 keV). On the other hand, one must bear in mind that in this case no additional density measurements or determinations of the geometric dimensions of the plasma are required.

1

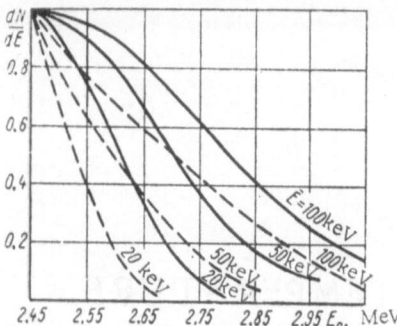

Figure 1. Form of the neutron spectrum for a Maxwellian energy distribution of the ions (solid curves) and for a monoenergetic plasma (dashed curves) for E = 20 keV, 50 keV, and 100 keV (the curves are normalized for $E = \overline{E}$).

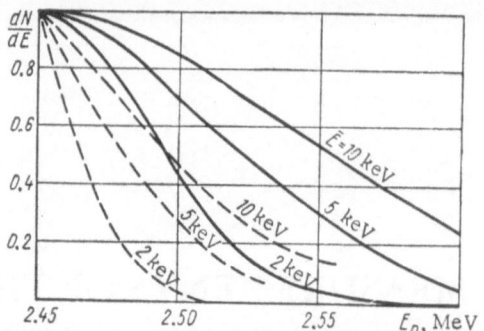

Figure 2. Form of the neutron spectrum for a Maxwellian energy distribution of the ions (solid curves) and for a monoenergetic plasma (dashed curves) at $\overline{E}$ = 2 keV, 5 keV, and 10 keV (the curves are normalized for $E = \overline{E}$).

The use of this method for plasma diagnostics has been discussed in several papers. It was possible to use the form of the n spectrum to distinguish between "true" neutrons emitted from the volume and "wall" neutrons generated in the chamber wall material when the walls are bombarded by accelerated ions [4-6]. This method of plasma diagnostics was also used for estimating the ion temperature in the Scilla device (under the assumption that the ion spectrum is characterized by a Maxwell distribution) [7].

In order to employ this method on a larger scale, several methodological problems must be solved. First of all, the real ion spectrum in a plasma is rather often not Maxwellian, and it is therefore necessary to have a relation between the observed n spectrum and the ion spectrum. The neutron spectrum corresponding to an equilibrium plasma has been calculated in the paper by Faust and Harris [8]. Another limit case for a monoenergetic plasma $f(E) = \delta(E - \overline{E})$ was considered in [9]. The dN/dE_n curves for the energies 2, 5, 10, 20, 50, and 100 keV are shown in Figures 1 and 2. One can assume that the n spectra corresponding to other types of energy distributions of the ions must be situated between these two types of curves. The simplest method of determining the ion temperature resides in this case on measurements of the spectral half-width, disregarding the form of the spectrum. It follows from the figures that this method can provide an accuracy of ±30% in practically the entire energy region under consideration (i.e., this method is much better than the integral method). The accuracy of the method can be substantially increased when the curves of the n spectrum are appropriately evaluated with mathematical methods in order to restore the ion spectrum (similar to the reconstruction of electron spectra from the form of x-ray spectra [10]).

It is also possible to determine $\overline{E}$ from other characteristic parameters (e.g., from $(d^2N/dE_n^2)_{max}$, i.e., from the maximum inclination angle of the contour curve), rather than from the half-width of the n spectrum.

The accuracy of the spectrometric neutron measurements is strongly affected by the "wall" effects. The cross section of elastic scattering at the nuclei of the wall materials is of the order of several barns, whereas the maximum energy loss of the neutrons amounts to

$$\Delta E_n = E_n \frac{4A}{(A+1)^2}$$ (for iron, we have $\Delta E_n/E_n = 7\%$). Single events of neutron scattering at nuclei constitute the principal mechanism resulting in neutron energy losses. In these processes, a considerable portion of the neutrons passes through the walls without collisions.

The distortions of the n spectrum can be insignificant at a wall thickness of less than 1 cm between the plasma and the detector and when a neutron collimator is used. When these conditions cannot be satisfied, it becomes necessary to make a preliminary calibration of the apparatus with the aid of a special source of monoenergetic neutrons, which is inserted into the working region of the plasma device. The results of the preliminary calibration can be used to establish the form of the n spectrum.

Methods involving the reaction

$$n + \text{He}^3 \rightarrow \text{T} + p + 0.765 \text{ MeV}$$

are obviously most appropriate for the spectrometry of fast neutrons.

A mixture of He^3 and heavy inert gases is used in proportional counters for spectrometric purposes. According to the literature data, the best counters can provide an energy resolution of up to 3% (average value 5%). The dead time is as long as $(2\text{-}5) \cdot 10^{-5}$ sec, and the counting efficiency amounts to 1% [11].

Another recording method employs a gas scintillator filled with He^3 and a 10% xenon admixture (total pressure 50 atm). The energy resolution is not better than 10-12% in this case, and the de-excitation time amounts to about 10^{-8} sec [12].

Finally, the above reaction can be employed in a diffusion chamber which is characterized by an energy resolution of 2%. However, since the time elapsing until the appearance of the tracks amounts to 0.05 sec, this method, as well as the method involving proportional counters, is not very suitable for observations of plasmas whose temperature changes rapidly in time. However, the latter methods are very appropriate for investigations of a stationary hot plasma.

REFERENCES

1. P. K. Toneman et al., in: "Controlled Thermonuclear Fusion," Moscow, Atomizdat (1958), p. 10.
2. L. A. Allen et al., ibid, p. 23.
3. L. A. Artsimovich, "Controlled Thermonuclear Reactions" [in Russian], Moscow, Fizmatgiz (1963).
4. B. Rose et al., Nature, 181:1630 (1958).
5. D. S. Hagerman and J. U. Meiser, in: "Controlled Thermonuclear Fusion," Moscow, Atomizdat (1958), p. 35.
6. R. A. Coombe and B. A. Ward, J. Nucl. Energy, 5(5):273 (1963).
7. K. Boyer et al., Phys. Rev., 119(3):831 (1960).
8. W. Faust and E. G. Harris, Nuclear Fusion, 1:62 (1960).
9. S. P. Bogdanov and V. I. Volosov, Zh. Tekhn. Fiz., Vol. 4 (1968).
10. M. S. Sodha et al., Progr. Theoret. Phys., 22(3):461 (1959).
11. "Fast Neutron Physics," J. Marion and J. Fowler, eds., Russian translation, Moscow, Gosatomizdat (1963), Vol. 1, p. 191.
12. S. A. Baldin and V. V. Matveev, Prib. i Tekhn. Eksperim., No. 1, p. 130 (1961).

ABSOLUTE SCINTILLATION MEASUREMENTS
OF NEUTRON PULSES GENERATED IN A HOT PLASMA

V. R. Lazarenko

Fluxes of fast neutrons generated in a hot plasma are characterized by particular properties. These fluxes are generated within a short time interval, and the repetition rate of the neutron pulses is usually very low. The number of neutrons in a pulse can vary from unity to 10^{10}-10^{14} neutrons, depending upon the type of the plasma device. The neutron energy is determined by the energy of the interaction between deuterons or between deuterons and titanium. Thus, the neutron counting method involving a piece-by-piece neutron count can seldom be used in plasma studies, and one has to resort to integral methods. Measurements of neutron pulses involving the activation of silver or indium and the ensuing counting of the number of decays of radioactive isotopes formed belong to the most widely employed techniques. This method is characterized by the disadvantage that the background resulting from the accumulation of long-lived isotopes increases slowly in time; apart from this, it is necessary to use neutron fluxes of a known intensity and with an appropriate neutron energy spectrum for calibration purposes.

In the case of sufficiently powerful neutron pulses, it is convenient to use scintillation counters recording knock-on protons in a hydrogen-containing scintillator for intensity measurements. The current pulse of the scintillation counter is recorded on an oscilloscope, and the number of neutrons in the pulse is derived from the area covered by the pulse form. This method permits one to measure the total number of neutrons as well as the time dependence of the neutron emission. The results obtained with this method agree to within 30-50% with the results of measurements based on indium activation.

A standard plastic prepared from polystyrene with additions of terphenyl and POPOP was used as the scintillator (diameter 45 mm). The thickness of 45 mm is suitable for reducing the contribution of repeated collisions between neutrons and protons to insignificant values. At a neutron energy of 2.4 MeV, the double collisions contribute about 2% in a plastic counter with a thickness of 1 mm, whereas they contribute 9% at a thickness of 5 mm, and about 25% at a thickness of 16 mm. The experimental conditions necessitated the use of a scintillator with a thickness of 16 mm.

In a collision with a proton, a fast neutron transfers about 50% of its energy to the proton, and thus, the knock-on proton has an average kinetic energy of about 1.2 MeV. The scintillator, in which the knock-on process occurred, delivers a light pulse of corresponding intensity. When the integral of the current pulse corresponding to the proton energy of 1.2 MeV is known at the photomultiplier output and when one has information on the cross section of neutrons scattered at hydrogen, one can determine the number of neutrons which passed through the scintillator.

4

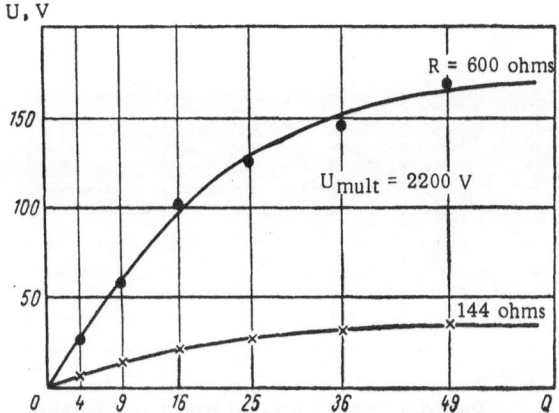

Figure 1. Pulse amplitude U at the output of the photomultiplier as a function of the intensity Q of the light flash for various load resistances.

One of four 13B multipliers was selected for the instrument. The current characteristics of the multipliers were used as the selection parameter. In all the multipliers tested, the linearity of the output signal was maintained to currents of 120–130 mA at a supply voltage of 2200 V. A xenon flash lamp provided with an exchangeable diaphragm for calibration purposes and with two light scatterers made of Teflon was used as the light source for the selection process. The pulse amplitude was recorded with a static voltmeter connected into the anode circuit of the multiplier via a diode. Load resistors of 144 ohms and 600 ohms were used (see Figure 1). In the linear region, the resulting current characteristics were independent of the load resistance of the photomultiplier (Figure 2).

The photomultiplier was calibrated from the current integral according to three different methods which provided results in close agreement. The gamma radiation of a sample of radioactive Cs[137] was used as the source. The accurate form of the photomultiplier pulse was determined in the first method and after that, with the relation between the pulse amplitude and the electron energy established, the quantity

$$q = \int\limits_{0}^{\infty} I\,dt = \frac{1}{R_{\text{load}}} \int\limits_{0}^{\infty} u(t)\,dt$$

could be calculated.

In this method, a 2 kohm resistor was used as the photomultiplier load, and the pulse shape was determined by successive measurements of the amplitude of the pulse reflected from a short-circuited delay line consisting of a cable of variable length (Figure 3). The measurements were made with a discriminator whose threshold was adjusted every time so that the counting rate remained constant, i.e., so that always the same region of the Compton electron spectrum was recorded. A multichannel amplitude analyzer was used for the energy measurements.

The second method made use of the fact that $q = U_{\max} C$ (with $U_{\max}$ denoting the pulse amplitude expressed in volts), provided that the RC constant of the multiplier output circuit is much greater than the de-excitation time of the scintillator ($\tau = 4.3 \cdot 10^{-9}$ sec). The load resistor was shunted with a capacitor so that the total capacitance of the photomultiplier output and of the ensuing input stage amounted to 695 ± 5 nF. Then the electron spectrum was recorded with an amplitude analyzer.

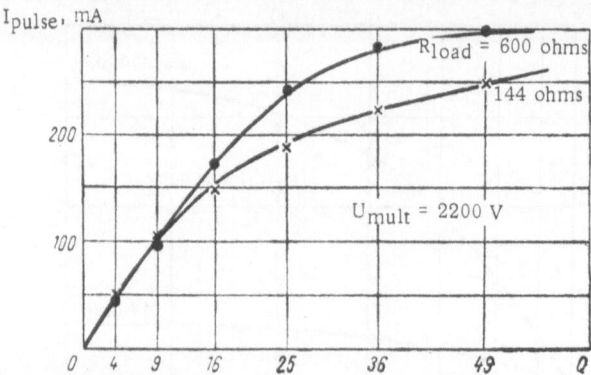

Figure 2. Pulse-current characteristics of the
photomultiplier at various load resistances.

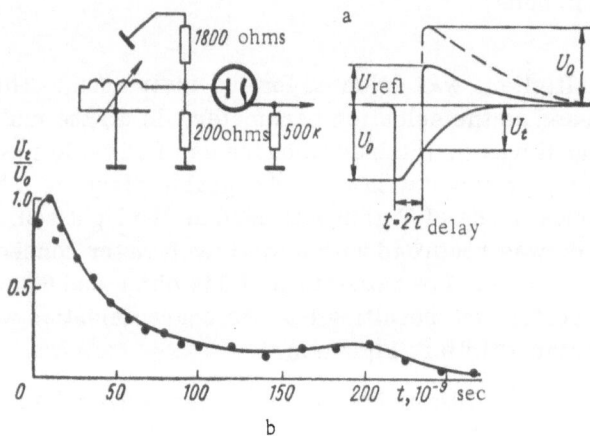

Figure 3. Schematic for determining the shape
of the scintillation pulse at the photomultiplier
output with the aid of a short-circuited delay
line (a); pulse shape obtained with the method (b).

Finally, $\int_0^\infty i\,dt$ can be determined when one assumes that the time dependence of the photomultiplier current is exactly exponential:

$$i = I_0 e^{-\frac{t}{\tau}}.$$

In this case, $q = I_0 \tau$, and the shape of the current pulse is given by the expression

$$U(t) = I_0 \frac{R\tau}{RC - \tau}\left[e^{-\frac{t}{RC}} - e^{-\frac{t}{\tau}}\right].$$

We obtain a relation between the amplitudes of the pulse current and the pulse voltage:

$$I_0 = \frac{U_m}{R}\left(\frac{RC}{\tau}\right)^{\frac{RC}{RC-\tau}}.$$

The quantities U_m, R, and C in this expression are measured, and τ is known. All these three methods of measuring $\int_0^\infty idt$ gave almost the same result: $1.29 \cdot 10^{-10}$, $1.39 \cdot 10^{-10}$, and $1.31 \cdot 10^{-10}$ C at an electron energy of 0.478 MeV. Thus, the average charge which a single electron supplies to the pulse is

$$q_{el} = 2.78 \cdot 10^{-10} \, C/MeV \pm 4\%.$$

When one switches to the knock-on protons, this value must be reduced by the factor 2.7 [1], and thus, a knock-on proton generated in the scintillator by a single neutron with an energy of 2.4 MeV results in an output pulse corresponding to the charge $q_n = 1.03 \cdot 10^{-10}$ C at the photomultiplier output.

Neutron pulses were measured with the LV device in which neutrons were generated at the time of plasma compression (pinch effect). The scintillation counter was placed at a distance of 15 m from the plasma device. This positioning allowed a time separation of the neutron pulses from the gamma and x-ray radiation pulses. The multiplier output was connected to an amplifier via a 50-m-long RK-50 cable to which a resistor of 150 ohms was connected as the load. The amplifier compensated for the signal attenuation in the cable and provided an additional amplification. The amplified signal was fed to the plates of an OK-21 oscilloscope and photographically recorded. The entire setup was calibrated with a GI-2M pulse generator.

The quantity $\int_0^\infty idt$ was graphically determined from the oscillogram, and the number of neutrons recorded by the instrument setup was obtained from the known q_n value:

$$N = \frac{1}{q_n} \int_0^\infty idt.$$

Since the absorption coefficient for neutrons with an energy of 2.4 MeV is $n\sigma = 0.126$ cm^{-1} for polystyrene [2], the efficiency of the neutron recordings with the 16-mm-thick scintillator was $\eta = n\sigma l = 0.2$. Taking into account the geometry of the experiment, a coefficient $\omega = S/4\pi R^2 = 5.6 \cdot 10^{-7}$ is obtained, where S denotes the scintillator area facing the neutron source, and R is the distance to the neutron source. The total recording efficiency is $\Omega = \eta\omega = 1.1 \cdot 10^{-7}$. During a series of measurements, this method was used to record neutron pulses with $5 \cdot 10^8$-$5 \cdot 10^9$ neutrons emitted during times of 0.25-0.35 mC/sec. The number of neutrons recorded by the scintillation counter amounted to about 100 with a total of about 500 neutrons incident on the scintillator.

Thus, the method is sufficiently sensitive for measuring weaker fluxes with less than 10^9 neutrons per pulse. The main difficulty which is encountered in this case results from the fact that it is necessary to take into account the background of gamma and x-ray radiation unless the experimental conditions permit one to separate these types of radiation in time. The background can be estimated by eliminating the fast neutrons which can be effectively slowed down by paraffin.

In conclusion, I express my gratitude to E. N. Lazarev and L. G. Tokareva for active participation in this work.

REFERENCES

1.   G. T. Wright, Phys. Rev., 91:1282 (1953).
2.   "Experimental Nuclear Physics," E. Segré, ed., Moscow, Foreign Literature Press (1955), Vol. 2, p. 199.

SEMICONDUCTOR DETECTORS FOR RECORDING FLUXES OF HYDROGEN IONS WITH ENERGIES OF 1—15 KEV

Yu. S. Maksimov

Semiconductor detectors with electron—hole junctions have been widely employed in modern experimental nuclear physics. These detectors possess several properties which recommend their application in plasma research. Among these properties, there are the relatively low energy required for generating an electron—hole pair (ε_{Si} = 3.6 eV and ε_{Ge} = 2.95 eV), the insensitivity to magnetic fields as strong as $5 \cdot 10^4$ Oe (at a temperature of about 300°K), the selective sensitivity to various types of radiation, the low inertia (time required for charge collection about 10^{-9}-10^{-7} sec), and the extremely small size. In order to establish the operation conditions and the adaptation of those detectors to plasma radiation studies, special experiments were made.

In the majority of cases [2, 3, 4], the energy range of the electrons, ions, and neutral atoms in experimental plasma devices extends from $0.5 \cdot 10^2$ to $0.5 \cdot 10^6$ eV. These energy values result in specific requirements for the semiconductor counters, i.e., for both the semiconductor materials and the corresponding production methods, and, in addition, for the electronic equipment used. In view of the energy range which in the case of particle radiation from a plasma comprises lower energies than in nuclear physics, the detectors must have a "thin window," i.e., a layer which is close to the surface and consists of an oxide film and a metal front contact. An effective charge collection does not take place in this layer.

General Conditions for Obtaining High-Energy

Resolution

A very high signal-to-noise ratio must be obtained by a considerable reduction of the intrinsic noise level of both the detector and the amplifier. The (mean square) value of the fluctuation in the number of charge-carrier pairs generated by a monoenergetic particle source can be represented in the form

$$\delta_N = \sqrt{FE\varepsilon}, \tag{1}$$

where F denotes the Fano factor, E the particle energy, and ε the average energy required for generating an electron—hole pair.

The energy resolution of a spectrometer operating with a semiconductor counter having a p—n junction as the sensitive element is given by the following quantities: statistics of the charge-carrier pair generation, Fano factor, intrinsic noise of the detector, and amplifier noise:

$$\Delta^2 = \Delta_N^2 F + \Delta_{det}^2 + \Delta_{ampl}^2. \tag{2}$$

The average energy dissipated in the generation of one electron—hole pair determines both the number N of charge carriers generated in the crystal during the de-excitation process and their statistical spread which depends upon the efficiency of the charge collection. The efficiency of the charge collection can be approximated by the formula

$$\eta \simeq \frac{\mu\tau_0 E}{x}\left(1 - e^{-\frac{x}{\mu\tau_0 \varepsilon}}\right), \tag{3}$$

where μ denotes the charge-carrier mobility, τ_0 the lifetime of the charge carriers, E the electric field in the p—n junction region, and x the depth of the region in which a depletion of charge carriers takes place.

Equations (1) and (2) permit one to determine the Fano factor when the relation between the resolution and the energy is known:

$$F = \frac{\Delta^2 - \Delta_{d,a}^2}{5.6 E \varepsilon}. \tag{4}$$

where Δ^2 denotes the total width at half the height of the spectral distribution function resulting from a monoenergetic particle source; $\Delta_{d,a}^2$ denotes the total width at half the height for a reference pulse which comprises the noise of both the detector and the amplifier; in the case of silicon and germanium, the Fano factors calculated with Eq. (4) [5] and the values stated in the papers by Blankenship [6] and Hansen [7] are $F_{Si} \leq 0.5$ and $F_{Ge} \leq 0.3$, respectively. The intrinsic noise of a semiconductor detector with a p—n junction is strongly affected by the leakage currents flowing through the junction; these currents depend upon the generation rate of the majority carriers in the depletion region. The contribution to the leakage current generated by the diffusion of minority charge carriers into the depletion region is much smaller than the principal component in real detectors.

The noise resulting from the fluctuations of these components of the leakage current is termed shot noise; the frequency spectrum of this kind of noise is uniform. When we disregard the effect of surface phenomena, the reverse saturation current of a counter diode with linear geometry is rather accurately given by the expression

$$I = A\frac{b\sigma_i^2}{(1 + b)^2} \cdot \frac{kT}{q}\left(\frac{\rho_n}{\sqrt{\tau_p}} + \frac{\rho_p}{\sqrt{\tau_n}}\right), \tag{5}$$

where d denotes the Boltzmann constant ($1.38 \cdot 10^{-23}$ W · sec · °K^{-1}), T the absolute temperature (°K), b the ratio of electron mobility to hole mobility in the semiconductor (μ_n/μ_p), σ_i the intrinsic conductivity of the semiconductor (ohm^{-1} · cm^{-1}), ρ_n the specific resistivity of the n region (ohm · cm), τ_p the lifetime of the holes in the n region (sec), ρ_p the specific resistivity of the p region (ohm · cm), and τ_n the lifetime of the electrons in the p region (sec).

Thus, a semiconductor material for the production of spectrometric counters with p—n junctions must be characterized by a very long lifetime of the charge carriers ($\tau_{Si} \geq 500$ μsec), and the specific resistivity must have the optimum value ($\rho_{Si} \sim 5 \cdot 10^2$-$5 \cdot 10^3$ ohm · cm).

Surface-Barrier Electron—Hole Junction

on n and p Silicon

In the case of a surface-barrier junction on silicon ($n_i \sim 10^{-10}$ cm^{-3}), the expressions for the depth of the depletion region (region in which the output signal of the counter depends linearly upon the energy of the particles to be detected) can be represented in the form

$$x_{Sl-p} = 0.32 \sqrt{\rho (U_0 + U)};$$
$$x_{Sl-n} = 0.53 \sqrt{\rho (U_0 + U)}, \tag{6}$$

where U_0 denotes the interval voltage at the p−n junction, i.e., the height of the surface barrier ($U_{0Si} \sim 0.5$ V); and U denotes the external reverse voltage at the counter (U = 0-10^3 V). A high-resistivity semiconductor makes it possible to vary the depth of the transition region within wide limits (x = 10-10^3 μ); thus, the surface-barrier counter can be made selective when particles (ions or neutral atoms) are recorded on an x-ray background. The width of the sensitive region of devices with diffused or drift junctions depends strongly upon the concentration distribution of the alloying admixture. Surface-barrier detectors have certain advantages over diffused junctions and drift junctions. Thus, the electron−hole junction begins practically at the surface (the metal contact on the front side may have a negligible thickness of 10 μg $\cdot$ cm^{-2}), and the nonequilibrium charge carriers, which were generated by the particle to be recorded, enter immediately into a region with a strong electric field. Apart from this, the production of surface-barrier counters does not involve thermal annealing or other technological processes which could lead to lattice defects detrimental to the counter operation.

It is generally accepted [8] that at a pure etched surface of a crystal there exist surface states, namely the Tamm energy levels, which result from the disturbed lattice periodicity at the boundary. These states are free bonds (quantum states not filled with electrons), which under certain conditions can act as traps for electrons arriving at the semiconductor surface from the bulk of the material. Thus, when an excess of electrons exists in the volume (this is usually the case in n type semiconductors), the charge of the surface states is negative. The electric quasineutrality in the near-surface layer is established by a space charge consisting of holes. This, in turn, results in an electric field directed from the bulk of the material toward the surface. A potential barrier for electrons develops, and the energy bands near the surface are upward bound, which means that a p type inversion layer has been formed (Figure 1a). The converse effect occurs in a hole-type semiconductor, i.e., when an excess of holes is present in the bulk of the material. The density of the surface states can reach values (when the material is treated with extremely pure chemical agents) which are determined by the number of the surface atoms, provided that a small fraction of the surface atoms captures electrons and that the charged state changes. The overall charge of the surface states is positive, and the same mechanism will remain effective, except for the fact that an electric field of reverse direction is generated and that the energy bands are bent downward near the surface (see Figure 1b). However, under real conditions, an oxide film is always present on any silicon surface; this film has a thickness ranging from a monomolecular layer to 10^4 Å and more, depending upon the production method. The binding energy of the valence electron which is considered bound to the nucleus of the atom (taking into account the shielding of the electrons of the outer shells) is given by

$$W_A \sim \frac{z_A^*}{\varepsilon n_A^2}, \tag{7}$$

with an accuracy determined by a coefficient. The notation is interpreted as follows: $z_A^* = z - \sigma$ denotes the effective charge of the nucleus; z is the charge of the nucleus corresponding to the atomic number of the element (σ denotes the shielding coefficient); ε is the dielectric constant; and n_A denotes the principal quantum number. $W_{Si} = 0.46$ relative units, and the binding energy calculated with the above formula for a compound is $W_{SiO_2} = 0.81$ relative units. According to the theory of chemisorption, particles with $W_{chem sorp} > W$ of the substrate attract electrons captured by surface states, and a charge redistribution takes place. Obvi-

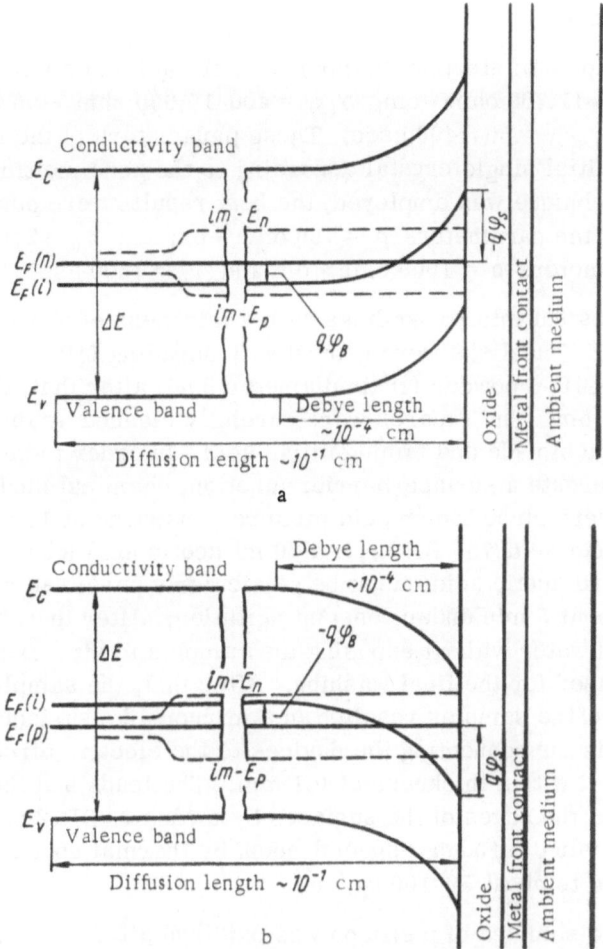

Figure 1. Energy diagrams of the surface region of (a)
an n type semiconductor and (b) a p type semiconductor. E_c, bottom of the conductivity band; E_V, top of
the conductivity band; φ_B, potential in the bulk of the
material; φ_s, surface potential; q, charge of a carrier;
ΔE, width of the forbidden zone; $E_F(i)$, Fermi level of the
material with intrinsic conductivity; $E_F(n)$ and $E_F(p)$,
Fermi levels of materials with electron and hole conductivity, respectively.

ously, this reduces the absolute value of the surface potential $|\varphi_s|$ of a junction on the surface
of n silicon. The reverse portion of the voltage−current characteristic of the counter (diode)
is therefore unfavorably affected. In the case of a surface-barrier transition on p silicon, this
shifting of the electrons leads to a more clearly visible inversion layer ($|-q\varphi_s|$ increases),
and the bands near the surface are more strongly bent upward. The breakdown voltage increases, and the reverse saturation current decreases. The above-described possible surface
mechanisms on n and p silicon emphasize the complex role of the oxide film which is always
present on a silicon surface and forms a passivating element protecting the surface from the
surrounding medium.

Counter Production

We prepared samples of electron (n type) and hole (p type) silicon with various specific resistivities $\rho_{(n)}$ = 150-11,700 ohm · cm, $\rho_{(p)}$ = 400-15,000 ohm · cm and carrier lifetimes $\tau_{0(n)}$ = 80-2100 μsec, $\tau_{0(p)}$ = 50-1400 μsec. These parameters of the silicon are obtained by purification of the initial single crystal according to the zone melting technique without crucible. When this technique was employed, the best results were obtained with counters of n type silicon having the parameters ρ = 700 ohm · cm and τ_0 = 2100 μsec and of p-type silicon having the parameters ρ = 1000 ohm · cm and τ_0 = 1400 μsec.

A silicon ingot was cut into round disks with a thickness of about 1.5 mm (cutting along the {III} crystal plane). The disks were ground with abrading MP micropowders. The final grinding was done with MP-5 powder (grain diameter 5 μ); after that, the disks were cut into squares of about 6 × 6 mm. The squares were carefully cleaned by immersing them twice into boiling carbon tetrachloride and fuming nitric acid. In order to remove the mechanically disturbed layer and to create a surface-barrier junction, chemical surface polishing was employed. The samples were etched in an acid mixture consisting of 15 ml hydrofluoric acid (55% HF), 40-50 ml nitric acid (75% HNO_3), and 20 ml acetic acid (concentrated CH_3COOH). Both the hydrofluoric and acetic acids must be particularly pure. At room temperature, the etching time was about 5 min under constant agitation. After that, the samples were washed by addition of deionized water without exposing the samples to air. Deionized water distilled in a special flask was used for the final washing. After that, the samples were dried with filter paper. This treatment of the samples resulted in a mirror-like surface with a hardly noticeable rippling. The leads for connecting the devices to the electric circuitry were made of silver-coated copper foil with a thickness of 0.1 mm. The leads had the form of 5 × 50 mm strips and were glued to the edges of the surfaces by means of BF-2 cement. Gold electrodes (front contacts) were applied through a special mask by thermal sputtering in vacuum. The film thickness amounted to about 50-100 μg/cm².

The surface of the samples of p silicon was oxidized after the etching. An oxygen discharge under a vacuum-tight bell jar was used for the oxidation. The samples were placed on the lower electrode, i.e., the cathode. A voltage of 100-400 V was applied to the electrodes. The current density amounted to about $2 \cdot 10^{-5}$ A/cm², and the gas pressure to 0.05-0.1 mm Hg.

Testing of the Counters and
Results of Measurements

In order to test the operation of the counters produced, the reverse current was recorded as a function of the reverse voltage applied to the counter. In addition, the capacity of the junction was measured as a function of the voltage applied. The counter operation was tested with α particles and fluxes of hydrogen ions having energies of about 1-15 keV.

Figure 2 shows the relation between the reverse current and the reverse voltage. Curves 2-4 are the reverse branches of voltage—current characteristics averaged over several (5-7) counters. Enhanced oxidation of the surface of p silicon improved the reverse branch of the characteristic, i.e., the reverse current decreases, and the breakdown voltage increases. On the other hand, n type silicon samples, which had been treated in the same way, exhibited a deterioration of their properties.

Curve 1 (see Figure 2) corresponds to the reverse branch of the voltage—current characteristic of n type silicon counters with a protective ring, i.e., an electrode applied to the same potential (with respect to ground) as the front contact of the counter.

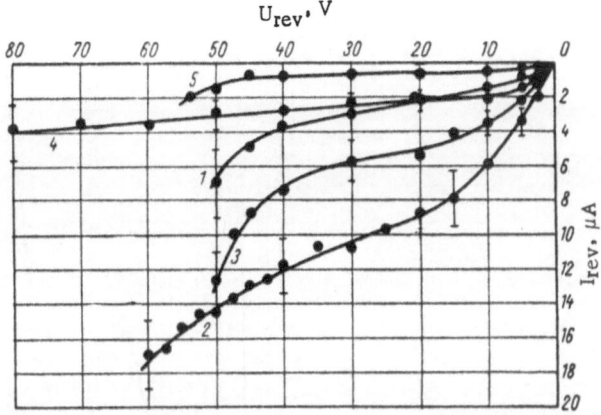

Figure 2. Relation between the reverse current I_{rev} and the reverse voltage U_{rev} for counters with surface-barrier junctions: 1) n type Si; ρ = 700 ohm · cm; τ = 2100 μsec; 2) the same with additional oxidation; 3) p type Si; ρ = 1000 ohm · cm; τ = 1200 μsec; 4) the same with additional oxidation; 5) n type Si; ρ = 700 ohm · cm; τ = 2100 μsec; tested with a protective ring.

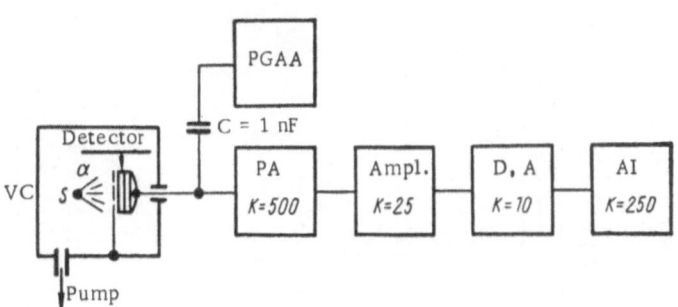

Figure 3. Block diagram of the α spectrometer (PGAA denotes the generator of pulses with accurately known amplitudes).

Figure 3 is the block diagram of the α spectrometer comprising a surface-barrier counter as the detector; a charge-sensitive preamplifier (PA, gain = 500) in the form of a cascade circuit with a 6S15P tube in the input stage; a linear amplifier (Ampl., gain = 25); an expander (D, A, gain = 10) comprising a discriminator D and an amplifier A; and, finally, a 100-channel amplitude analyzer (type AI-100-1). A similar block diagram was described in [9]. The α particle source was placed in the vacuum chamber VC; the measurements were made at room temperature. Figure 4a shows the resolving power of the semiconductor α spectrometer. Groups of the fine structure of the α radiation emitted by Pu^{238} and Cm^{242} appeared clearly in the observed spectra. One must bear in mind that the energy resolution of the detector itself is better than the resolution shown in the figure. The reason is that the noise level of the preamplifier amounts to about 14-16 keV at an external input capacity C = 100 nF. Thus, the preamplifier contributed considerably to the half-width of the peak generated by the α particles, as can be seen in Figure 4b. In order to generate a reference peak, pulses of a generator with accurately known amplitude (PGAA) were fed through a capacitor C = 1 nF to the input of the preamplifier and adjusted so that they had the same amplitude as

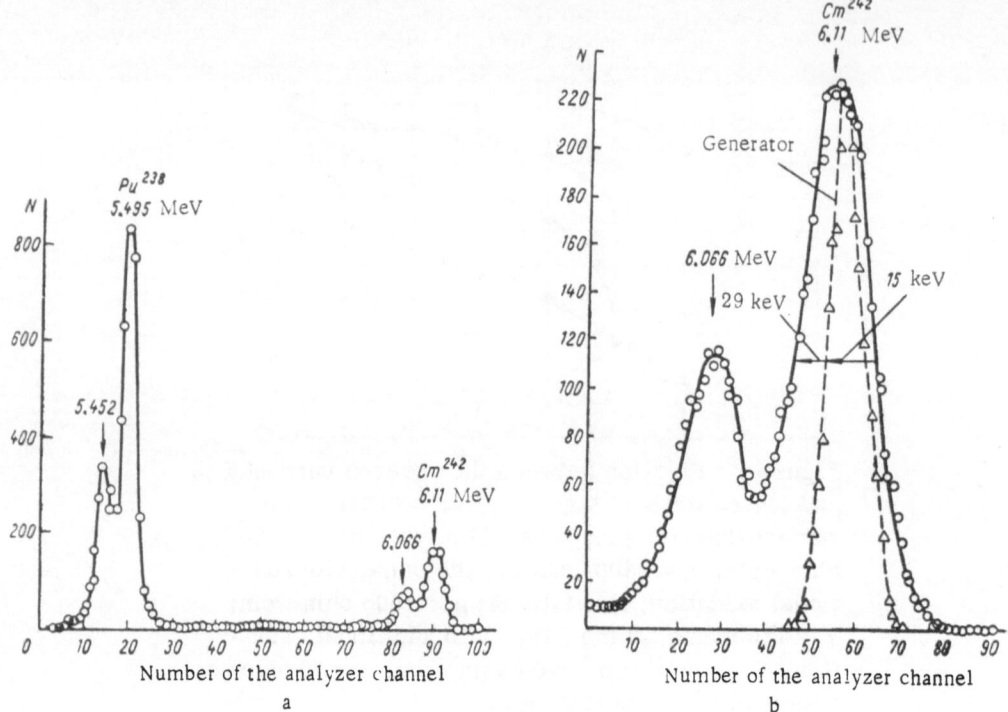

Figure 4. Resolving power of the α spectrometer.

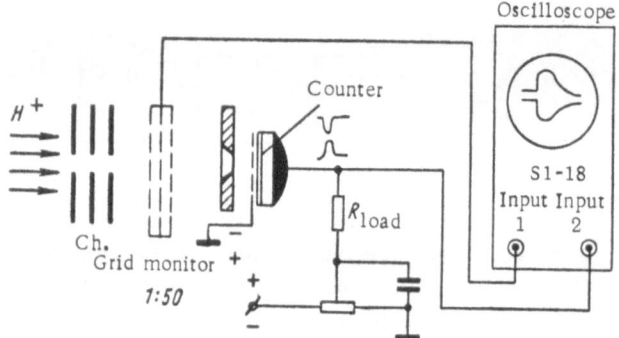

Figure 5. Connection of the counters for recording
the flux of hydrogen ions.

the pulses which were generated by the Cm^{242} α particles and derived from the load resistor of the counter. Then the α source was removed while the counter remained switched on ($C_{counter} \simeq 100$ nF). It was assumed that the spread in the amplitude of the pulses provided by the generator was negligible during the time of the measurements. Optimum energy resolution can be obtained by selecting the optimum values for the RC constant for the differentiating and integrating circuits.

  Figure 5 shows the testing circuit for recording fluxes of hydrogen ions. The influence of secondary electron emission was eliminated by applying a magnetic field to the detector and the grid monitor. Figure 6 shows the dependence of the counter-output signal on the intensity of the particle flux for various particle energies. At low densities of the ion fluxes, a linear relation is observed for all energies. The slope of the linear curve sections increases with increasing particle energy. The nonlinear curve section at very large fluxes may result from

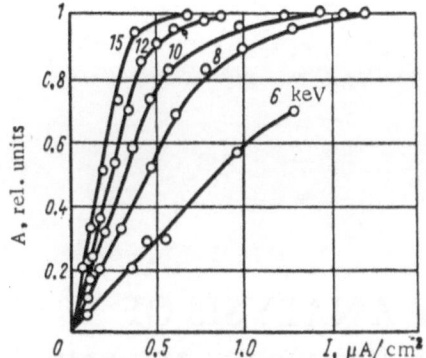

Figure 6. Output signal of the counter as a function of the flux intensity of the ions for various energies.

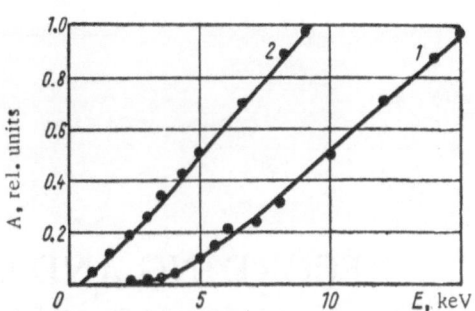

Figure 7. Output signal of the counter as a function of the energy of atomic ions.

the "plasma effect" in the depletion region of the counter. This effect limits the current generated by the recorded particles. An estimate shows that for the counters used in our work, the upper limit of the ion-flux intensity must be less than

$$I \leqslant 5 \cdot 10^{-3} \frac{1}{E - E_{\text{loss}}} \text{A/cm}^2, \tag{8}$$

where E denotes the energy of the ions recorded (eV) and E_{loss} the energy lost in the gold layer (eV). Figure 7 (curve 1) shows the output signal of the counter as a function of the energy of the hydrogen ions in the energy range 1–15 keV for a constant current intensity and a constant voltage. Curve 2 takes into account the energy losses in the gold layer.

The surface-barrier counters described in this article were used [10] for recording fluxes of neutral atoms with energies of about 10 keV in a device for adiabatic plasma compression.

REFERENCES

1. G. Dearnley and A. B. Whitehead, Atomic Energy Res. Establishment, NR-3437, 34 (1960).
2. L. A. Artsimovich et al., Atomnaya Energiya, 3:84 (1956).
3. A. M. Andrianov et al., Proc. Soc. Intern. Conf. on Peaceful Uses of Atomic Energy, Geneva, 31:348 (1958).
4. B. G. Brezhnev and Yu. S. Maksimov, Proc. Fifth Intern. Conf. on Ioniz. Phen. in Gases, Munich, Vol. 2, p. 1319 (1961).
5. B. V. Fefilov and L. Kumpf, Low-Noise Pulse Amplifiers for Semiconductor Detectors, Joint Institute for Nuclear Research, Dubna (1965).
6. J. L. Blankenship and W. F. Mruk, Bull. Amer. Phys. Soc., Ser. II, 9(1), p. 49 (1964).
7. W. L. Hansen and B. V. Jarrett, Report UCRL-11589 (August 7, 1964).
8. I. E. Tamm, Sow. Phys., 1:733 (1932).
9. Yu. S. Maksimov et al., Prib. i Tekhn. Eksperim., No. 2 (1966).
10. A. V. Bortnikov et al., Atomnaya Energiya, 18:256 (1965).

RECORDING AND ENERGY ANALYSIS OF LOW-ENERGY H_1^+, H_2^+, AND He$^+$ IONS BY MEANS OF SURFACE-BARRIER SILICON COUNTERS

G. F. Bogdanov, M. M. Dremin, and B. P. Maksimenko

Surface-barrier silicon counters have opened new possibilities of recording and analyzing charged and neutral particles emitted from a plasma [1, 2]. In the majority of cases, the energies of these particles do not exceed several tens of kiloelectronvolts. The H_1^+-ion energy below which the probability for the formation of an electron—hole pair in silicon is negligibly small amounts to about 0.25 keV (about 1 keV for He^{++}) [3]. Therefore, in principle, the energy threshold of counters used for the spectrometric analysis of particles is determined by the losses in the insensitive counter "window" and by the statistical fluctuations of the number of electron—hole pairs created by a single particle. The energy threshold can be very low (1-2 keV for H_1^+ ions).

Surface-barrier silicon counters, as well as ionization chambers and scintillation counters, can be used in two ways, namely for spectrometric measurements and for flux measurements.

In [4] a linear relationship between the amplitude of the counting pulses and the energy of H_1^+ ions was obtained in the energy range between 18 and 250 keV when uncooled counters were used for spectrometric measurements. The total width at half the amplitude maximum amounted to 8.5 keV and was independent of the energy of the H_1^+ ions. It was impossible to measure lower energies, because the noise of the counter—preamplifier system was too strong.

In the flux measurements of [2], a linear relationship was obtained between the counter output signal and the energy of H_1^+ ions in the energy interval 1-15 keV. The measurements were made at a constant flux of the incident ions (of the order of 10^{-7} A/cm^2) and a constant voltage applied to the counter. It is difficult to use counters in flux measurements when the measured current, which is amplified by the amplification coefficient (gain) of the counter, is commensurable with the reverse counter current at the operating voltage applied. The reverse current can be reduced by cooling the counter. Another possibility is to operate the counter without applying a voltage, provided that the particle path terminates completely in the sensitive layer of the counter. In the case of silicon with a specific resistivity of about 150 ohm · cm and a barrier height of about 0.5 V, the width of the sensitive layer amounts to about 4.5 μ, i.e., the paths of H_1^+ ions with energies of up to 400 keV and He^{++} ions with energies of up to 1.7 MeV end in a layer of that thickness.

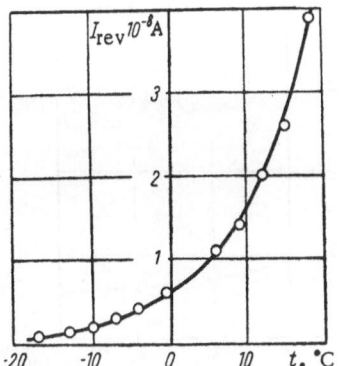

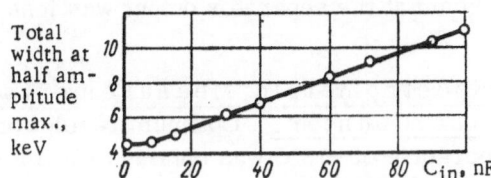

Figure 1. Relation between the total width at half the maximum amplitude and the external capacity at the preamplifier input.

Figure 2. Dependence of the reverse counter current upon the temperature; voltage applied to the counter, 50 V.

We measured in our work the spectrometric characteristics of counters by recording H_1^+, H_2^+, and He^+ ions as a function of the temperature. Furthermore, we investigated the relation between the amplification coefficient of counters operated in flux measurements without application of a voltage, on the one hand, and both the energy and the total flux of H_1^+ (recorded by the counter), on the other.

Counters and Preamplifier

Two sets of surface-barrier silicon counters with sensitive areas of about 3 and 10 mm^2 (for spectrometric and flux measurements, respectively) were produced as usual from n silicon having a specific resistivity of about 150 ohm · cm. The thickness of the gold coating amounted to about 20 and 40 μg/cm^2, respectively. At a voltage of 50 V and a temperature of +18°C, the reverse current was less than $4 \cdot 10^{-8}$ A for the first counter type and $2 \cdot 10^{-7}$ A for the second.

The preamplifier used in our work consisted of two amplifying stages.

The first stage consisted of a charge-sensitive circuit with an input stage using 6S15P tubes [5]. The amplification coefficient (gain) of the stage without inverse feedback was 440, and the feedback capacity amounted to 1 nF. This stage had been designed by V. G. Brovchenko.

The second amplifying stage had a gain of 80 and consisted of a usual three-tube circuit with negative feedback.

The rise time of the pulse front was less than 0.1 μsec in the preamplifier.

A set of differentiating and integrating circuits with the time constants 0.5, 0.7, 1.0, 1.5, 2.0, and 2.5 μsec was provided in the main amplifier. Optimum resolution was obtained with $\tau = \tau_{diff} = \tau_{int} = 0.7$ μsec.

Figure 1 shows the total width at half the maximum amplitude of the pulse spectrum at the preamplifier output as a function of the external capacity at the input (pulses supplied by a GI-4M generator were recorded). When the external capacity vanishes, the total width at half the maximum amplitude amounts to 4.5 keV instead of the calculated value 2.3 keV. This difference seems to result from improperly selected input tubes and imperfections in the construction.

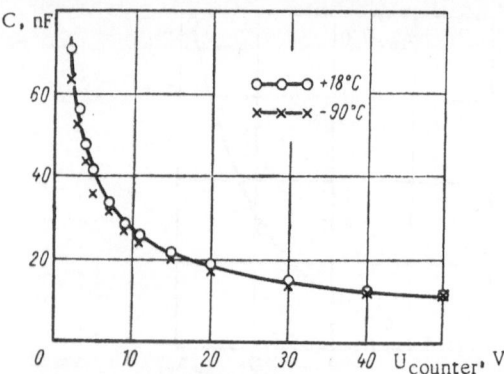

Figure 3. Voltage dependence of the
counter capacity.

Results of the Measurements

The measurements were made with a magnetic sector separator (60°) with beams of H_1^+, H_2^+, and He^+ ions. The accelerating voltage was varied from 0 to 100 kV. The energy spread of the ions in the beam at the separator output was less than 2.5%.

Spectrometric Measurements with the Counter.

Optimum resolution at a temperature of +18°C was obtained when a voltage of 40-60 V was applied to the counters. The resolution remained practically constant within this voltage range. Usually, the counters were operated with a voltage of 50 V.

Figure 2 shows the dependence of the reverse counter current upon the temperature when a voltage of 50 V was applied to the counters. The reverse current depends strongly upon the temperature and when the counter is cooled from +18°C to −17°C, the reverse current is reduced 40 times. For the cooling, the counter was placed in a vacuum chamber (pressure about $5 \cdot 10^{-5}$ torr). The temperature was measured with a copper−constantan thermocouple which was attached to the ohmic contact of the counter.

Figure 3 shows the dependence of the counter capacity upon the voltage applied to the counter. This dependence was measured at the temperatures +18°C and −90°C. Obviously, the capacity of a cooled counter is only slightly smaller than that of a warm counter when the voltages applied to the counters are small. When the voltages applied to the counters are high, the capacity is practically unchanged by the cooling, and this agrees with the results of [6].

The spectra were recorded with an AI-100 amplitude analyzer.

Figure 4 shows the amplitude characteristic of a counter cooled to the temperature −120°C; the amplitude characteristic was measured for three amplifier gains and in the following energy ranges: H_1^+ ions, 11-90 keV; H_2^+ ions, 13.5-90 keV; and He^+ ions, 14.5-90 keV. The linear relationship between the ion energy and the pulse amplitude is clearly visible, i.e., the energy spent in forming an electron−hole pair is independent of the ion energy in this energy range. The extrapolated straight lines intersect the energy axis at three points, and the corresponding "sections" are respectively equal to about 0.75 keV for H_1^+ ions, 1.5 keV for H_2^+ ions, and 3 keV for He^+ ions. The difference in the counter-pulse amplitudes resulting from H_1^+, H_2^+, and He^+ ions at some energy can be explained by different energy losses in the gold coating of the counter. When a correction for the energy losses of the H_1^+ ions in gold is introduced, the corresponding straight line passes through the coordinate origin, i.e., no substantial energy losses occur in the inversion layer on the silicon surface. The energy required for generating an electron−hole pair amounts to 3.5 ± 0.5 eV.

Figure 5 shows the amplitude distribution of the counter pulses when H_1^+ ions with the energies 45 and 16 keV were recorded. These measurements refer to a counter cooled to −120°C. The total width at half the maximum amplitude amounts to 5.5 keV for both energies. A slight broadening of the peak which is observed in the case of the H_1^+ ions with energies of 16 keV can be explained by the fact that the decrease in the amplitude of the pulses begins to become noticeable when the relative intensity of the noise pulses having comparable amplitudes increases. The total width at half the maximum amplitude amounts to 7.5 keV when the counter

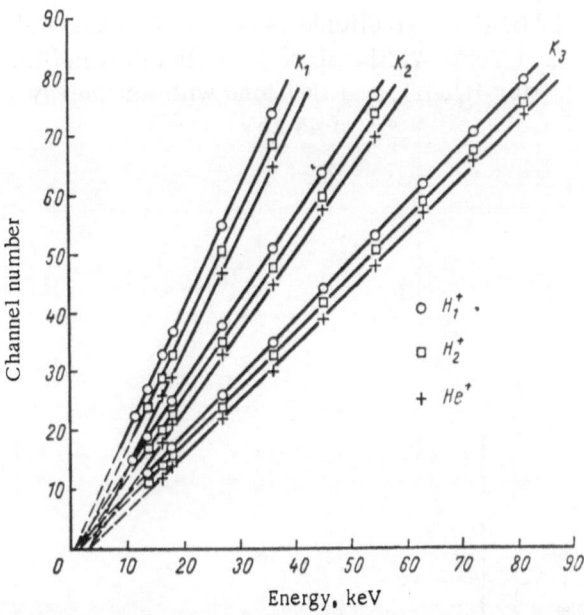

Figure 4. Relation between the channel number and the energy of H_1^+, H_2^+, and He$^+$ ions for three amplifiers gains $K_1 > K_2 > K_3$. Counter temperature, $-120°C$.

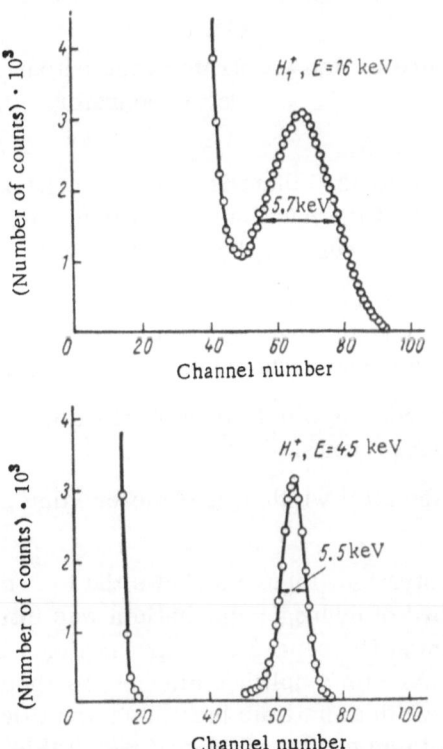

Figure 5. Amplitude distribution of the counter pulses for H_1^+ ions with the energies 16 and 45 keV. Counter temperature, $-120°C$.

TABLE 1. Amplitude (A) of the Counter Pulses
and Total Widths at Half Maximum Amplitude
for H_1^+, H_2^+, and He$^+$ Ions with an Energy
of 45 keV

Ions	H_1^+		H_2^+		He$^+$	
t, °C	—120	+18	—120	+18	—120	+18
A, rel. units	1	—	0.93	—	0.86	—
Total width at half maximum amplitude, keV	5.5	7.4	6.5	8.0	7.7	8.9

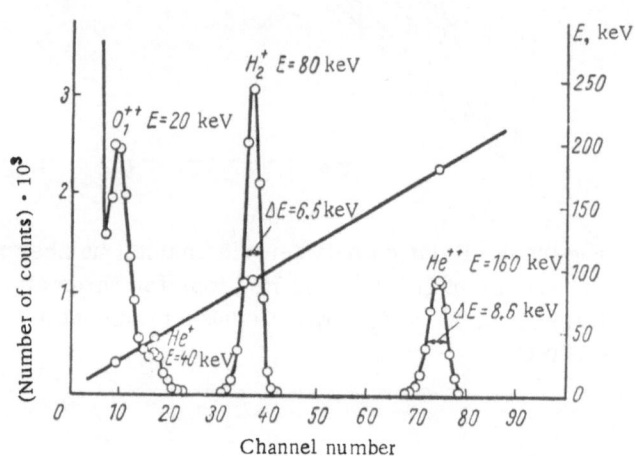

Figure 6. Ion spectrum at the output of the magnetic separator.

is operated without cooling, but under otherwise equal conditions (+18°C) (see Table 1). The decrease in the total width at half the maximum amplitude during cooling results mainly from a reduction of the reverse counter current, because the capacity of the counter is almost unchanged [5].

A slight decrease in the counter-pulse amplitude (about 1.5% for H_1^+ ions at an energy of 45 keV) was observed when the counter was cooled. This agrees with the results of [6].

Table 1 lists composite results which were obtained when H_1^+, H_2^+, and He$^+$ ions were recorded in spectrometric measurements.

According to Table 1, the total width at half the maximum amplitude increases with increasing particle mass.

For the purpose of illustration, Figure 6 shows the ion spectrum obtained with the magnetic separator when a mixture of hydrogen and helium was introduced into the source. When the energy spread of the beam at the separator output is taken into account (about 2.5%), the true total width at half the maximum amplitude amounts to about 7.7 keV for He^{++} ions. The agreement between the total width at half the maximum amplitude of He^{++} ions with an energy of 160 keV and of He$^+$ ions with an energy of 45 keV (see Table 1) indicates that the total width at half the maximum amplitude of helium ions is independent of the energy and the charge state of the ion appearing at the counter input. This result is explained by the fact that at energies of up to 500 keV, the equilibrium in the charge state of the helium ion—helium atom beam is established in layers as thin as 3 μg/cm^2 [7]; the thickness of the gold coating amounts to about 20 μg/cm^2.

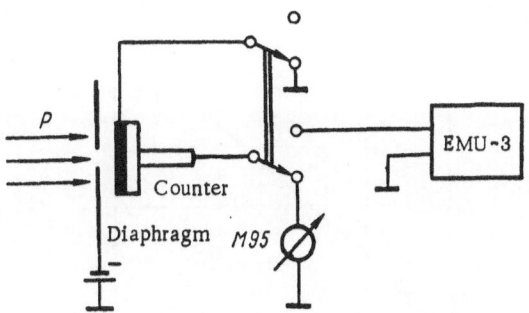

Figure 7. Connection of the counter to the
measuring instrument.

Flux Measurements with the Counters. In order to test the possible use of
the counters for measuring small fluxes (10^{-12}–10^{-13} A), we measured the relation between the
gain of a counter operated without a voltage and the energy of the H_1^+ ions or the total flux of H_1^+
ions incident per unit sensitive counter area.

The counter gain is defined as

$$K_{ampl} = \frac{I_{out}}{I_{in}} = \left[\frac{E_0 - \Delta E}{\varepsilon} \beta + 1 \right],$$

where I_{in} denotes the flux incident on the counter; I_{out} is the output current of the counter; E_0
denotes the energy of the incident ions; ΔE denotes the ion-energy loss in the insensitive "window" of the counter; ε is the energy required for the generation of an electron–hole pair; and
β is the yield coefficient for minority charge carriers.

Under the condition

$$\frac{E_0 - \Delta E}{\varepsilon} \gg 1, \quad \beta = 1$$

we obtain

$$K_{ampl} = \frac{E_0}{\varepsilon} \left(1 - \frac{\Delta E}{E_0} \right).$$

The connection of the counter to the measuring instrument is shown in Figure 7. A negative voltage was applied to a diaphragm placed before the counter in order to eliminate secondary electron emission from the gold coating of the counter. The flux incident on the counter
was measured with an electrometric EMU-3 amplifier, while one of the counter leads was dis-
connected. The counter output current was measured with an M-95 galvanometer.

Figure 8 shows the counter-gain coefficient as a function of the energy of the H_1^+ ions for
the energy range 6–36 keV and for an incident ion density of about 10^{-11}–10^{-12} A/cm^2. The ex-
trapolation of the straight line results in a "cutoff" at about 1.5 keV (thickness of the coating
about 40 μg/cm^2). When a correction for the energy losses of the H_1^+ ions in gold is introduced, the straight line passes through the coordinate origin. The energy for the formation
of an electron–hole pair (this energy can be determined from the slope of the straight line)
amounts to 3.6 ± 0.5 eV. Thus, the electric field of the intrinsic surface barrier of the counter
suffices for collecting the resulting minority charge carriers at the electrodes. It was a result stated in [2] that at currents of about 10^{-7} A/cm^2, the counter gain depended strongly upon
the voltage applied to the counter and increased with increasing voltage. Since in [2] the density of the incident flux was much greater than in our case (about 10^{-7} A/cm^2 versus 10^{-11} A/cm^2),

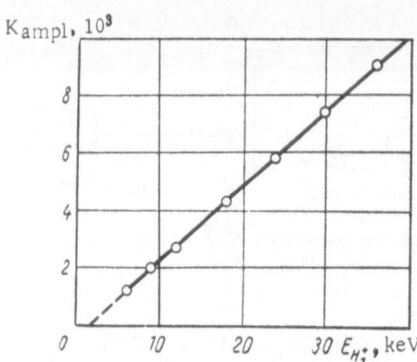

Figure 8. Relation between the counter-gain coefficient (no voltage applied to the counter) and the energy of the H_1^+ ions.

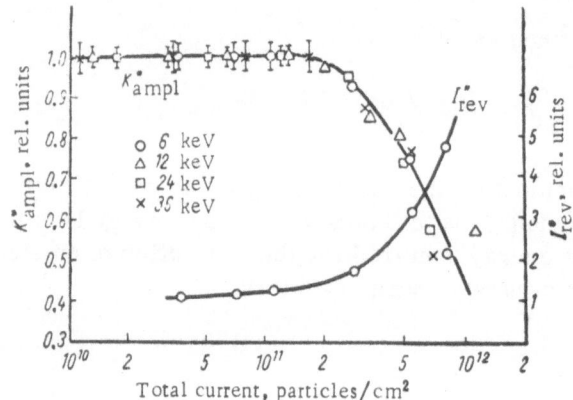

Figure 9. Dependence of the gain and the reverse counter current upon the total H_1^+-ion flux per unit sensitive counter area.

the decrease in the counter gain observed without a voltage applied to the counter can be explained by a high concentration of majority charge carriers in the sensitive counter layer. This results in a decrease in the electric field inside the region filled with minority charge carriers. The time required for collecting the minority charge carriers increases and this, in turn, increases the probability of losses of minority charge carriers by recombination.

Figure 9 shows the dependence of the normalized counter gain ($K_{ampl}^* = K_{ampl}/K_{ampl\,irr}$, where $K_{ampl\,irr}$ denotes the counter gain before the irradiation) and the reverse current through the counter (with 50 V applied to the counter; $I^* = I/I_0$, where I_0 denotes the reverse current through the counter before irradiation) upon the total flux of H_1^+ ions per unit sensitive counter area. The volt−ampere characteristics of the counter were measured in the time interval between irradiations. The counter gain remained constant, within the accuracy limits of the measurements, up to total H_1^+-ion fluxes of about $2 \cdot 10^{11}$ particles per cm^2. At higher flux densities, the counter gain began to decrease and a strong dependence upon the energy of the H_1^+ ions was observed. When the total flux increased, the reverse counter current increased continuously, but the slope of the volt−ampere characteristics remained almost unchanged. A noticeable increase in the reverse current was accompanied by a strong reduction of the counter gain.

TABLE 2

Medium	$t_1 = 3\tau_1$, min	$(I_{rev}/I_0)_{t_1}$	τ_2, min	t_2, h	τ, h	$(I_{rev}/I_0)_{t=20\,h}$
O_2	15	1.5	45	2	35	0.65
N_2	75	2	120	10	35	0.8
Vacuum $(10^{-1}$—10^{-2} torr)	135	2.7	—	30	35	1.5

Once the irradiation had been stopped, the counter gain was gradually restored, and the reverse counter current decreased to the original values. The rate of decrease in the reverse current and the restoring of the counter gain depended strongly upon the composition of the atmosphere in which the counter was located. O_2 and N_2 atmospheres were created by introducing oxygen ("for technical purposes" brand) and by evaporating liquid nitrogen into the evacuated setup. The gas in the setup was not analyzed.

Table 2 lists the reduction of the reverse counter current after the end of the irradiation.

The curves indicating the reduction of the reverse current after the end of the irradiation are characterized by a rather steep initial section with a constant decline τ_1, which depends upon the conditions in the chamber. On the curves which were obtained in an oxygen—nitrogen atmosphere, one can thereafter separate a section with a time constant τ_2. Finally, the ensuing decrease in the reverse current has the same form in oxygen, nitrogen, and vacuum (time constant τ_3). The curve depicting the gain restoration in vacuum is composed of two exponential curves. The τ_1, τ_2, and τ_3 values are listed in Table 2. Table 2 lists in addition the reverse currents I_{rev} referred to the reverse current I_0 before the irradiation at the times $t_1 = 3\tau_1$ and 20 h after the end of the irradiation. Moreover, Table 2 lists the time t_2 during which the reverse current reaches its initial value. The counter gain is reassumed much faster than the original reverse current.

It follows from these results that the counters used in spectrometers can be restored during a relatively short time after an overload. The restoration of the counters is accomplished by introducing into the setup a relatively inert gas, such as nitrogen. The restoring is very important when counters are used in plasma devices, because it is not always possible to introduce oxygen into plasma devices and it is not desirable to open these devices frequently and to bring them in contact with the outside atmosphere.

The strong dependence of the time required for restoring the counter upon the composition of the atmosphere around the counter indicates that the counter characteristics are affected not only by possible radiation damage, but also by surface phenomena which occur when the counters are bombarded by low-energy H_1^+ ions.

In conclusion, the authors express their gratitude to V. G. Brovchenko for designing the preamplifier and for his assistance, to G. M. Novikov for producing the counters, and to V. V. Strulev for assembling the preamplifier and for participating in the measurements.

REFERENCES

1. G. F. Bogdanov and B. P. Maksimenko, Atomnaya Energiya, 19:449 (1965).
2. N. N. Brevnov et al., Atomnaya Energiya, 20:149 (1966).

3.   F. Seitz, in: "Radiation Effects upon Semiconductors and Insulators," Russian translation, S. M. Ryvkin, ed., Moscow, Foreign Literature Press (1954).

4.   R. I. Ewing, IRE Trans., N.S., 9(3):207 (1962).

5.   V. G. Brovchenko and Yu. D. Molchanov, Prib. i Tekhn. Eksperim., No. 4, p. 5 (1964).

6.   F. Cappelani and G. Restelli, Nuclear Instr. and Methods, 25:2230 (1964).

7.   S. K. Allison, Rev. Mod. Phys., 30:1137 (1958).

DEVICE FOR PLASMA DIAGNOSTICS
WITH A MULTICOMPONENT BEAM
OF FAST NEUTRAL PARTICLES

N. I. Alinovskii, Yu. E. Nesterikhin, and B. K. Pakhtusov

An active method for the diagnostics of a high-temperature plasma by means of beams of fast neutral particles has been successfully tested in the last few years [1-5]. The method is very promising, does not involve a direct contact with the plasma, is characterized by high resolution in time and space, and enables plasma-parameter measurements in a range in which measurements with other techniques are difficult.

As has been shown, e.g., in [1], measurements of the relative attenuation of neutral particle beams passing through a plasma can be used to obtain the time-dependent pattern of the principal plasma parameters, namely n_1, the ion density; n_0, the density of the neutral component; and T_e, the electron temperature.

The present article describes a device for the diagnostics of a hydrogen plasma with a three-component neutral beam (H^0, H_2^0, and He^0) having an energy of several kiloelectronvolts. The simultaneous use of three beams increases both the accuracy and reliability of the results of a diagnostic test. The use of a beam of hydrogen atoms enables measurements of the ion density in a hydrogen plasma. Once the ion density is known, one can evaluate the electron temperature of the plasma from the relative attenuation of the helium beam. The attenuation of the beam of fast hydrogen molecules provides additional information on the electron temperature or, in the case of a low degree of plasma ionization, on the density of the neutral component.

Light atoms can be recommended for testing for several reasons. First, one can obtain a very high time resolution with light atoms, because the resolution is basically determined by the time of flight of the test particles in the plasma; second, light atoms facilitate ionization measurements in the neutral beam portion which was attenuated by interactions with the plasma, and thus, one considers stripping reactions in the gas because the corresponding cross section assumes its maximum in the case of light atoms; third, one can use a magnetic analyzer with a high mass dispersion which is also greatest for light particles [6].

Experimental Setup

Figure 1 shows schematically the setup described. A multicomponent beam of neutral particles with an energy of several kiloelectronvolts is generated in a neutral particle source 1 which comprises a usual high-frequency ion source 7 (f = 50-70 MHz; power of the discharge P = 100 W). The charge exchange of the ions in the source takes place within the channel of an accelerating electrode 3 (l = 60 mm, d = 2.5 mm) and involves the gas which flows in from the source. This arrangement made it possible to simplify the source and to reduce its size, because, first of all, it was no longer necessary to use a charge-exchange chamber and, hence, a huge vacuum system, and second, because it was possible to simplify the supply circuit for the high-frequency generator (the generator is on ground potential). In order to obtain a vacuum of about $5 \cdot 10^{-5}$ mm Hg in the volume following the accelerating electrode, a single TsVL-100 pump sufficed during the operation of the source.

At a distance of about 0.5 m from the source, the equivalent current of neutral hydrogen atoms (measured with the current of secondary electrons which are knocked from the retractable target 7 by the neutral atoms) reached 80 μA at a beam diameter of about 0.5 cm.

The electric field between the accelerating electrode 3 and the chamber walls stops the ions which do not participate in the charge-exchange process; this leads to a shielding of the neutral beam. A capacitor which deflects the ions in the transverse direction is provided for the same purpose. The neutral beam which was attenuated by the interaction with the plasma is once more separated from the charged particles which enter into the beam from the plasma, and passes into stripping chamber 13. This chamber has the form of a gas-delay line, i.e., it consists of a set of concentric stainless steel rings which are closely spaced (spacing about 2 mm). The internal ring diameter increases gradually toward the chamber outlet (from 6 to 14 mm) and creates an aperture angle of not less than 4-5° for viewing the center section of the plasma. The chamber has a length of 12 cm. The pressure is of the order $(1-5) \cdot 10^{-4}$ mm Hg. Air is used as the stripping gas.

The set of concentric rings is isolated from the housing by means of Teflon sleeves. By adjusting the potential applied to the stripping chamber, the ions generated by stripping reactions in the neutral beam can be transferred from one energy interval into another to which the electrostatic ion-energy analyzer 15 and the magnetic mass analyzer 18 are adjusted. This facilitates the adjustment of the entire system when one switches to a beam with a different energy.

The electrostatic ion-energy analyzer, in which a cylindrical capacitor [8] with an aperture angle of 127°17' is used, serves for separating particles with a certain, well-defined energy from the ion beam.

During the operation of the stripping chamber, the pressure in the analyzer amounts to $(3-6) \cdot 10^{-5}$ mm Hg.

The mass resolution of the monoenergetic ion beam takes place in the magnetic analyzer 18 which consists of sectors and has an aperture angle of 60°. A collector 20 for regulating the current of the ion beam leaving the electrostatic analyzer is provided in the chamber of the analyzer.

The ion beams are recorded with ion-electron converters 21. The dispersion of the magnetic analyzer amounts to about 43 mm per proton mass. Therefore, small-size 16B photomultipliers can be used without connecting light guides. The aperture of the target relative to the center of the plasma device is not less than 3°. In calculating the aperture angle, the focusing effect originating from the charged particles was taken into account. The focusing effect depends upon the angle of the cylindrical capacitor which has an aperture angle of 127°17' [8].

In order to simplify the adjustment of the entire system, a d.c. signal modulation of the multiplier photocathode is used; the modulating frequency is 500 kHz. The modulating high-frequency voltage has an amplitude of not more than 10-15 V and is applied to the photocathode rather than to a modulating grid which is usually provided for this purpose [9]. Application of

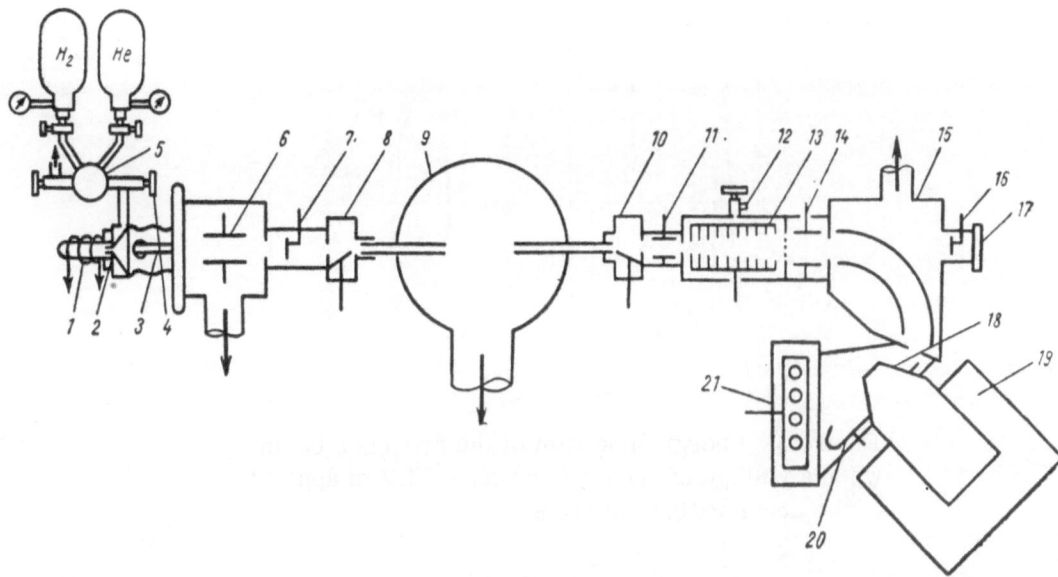

Figure 1. Scheme of the device. 1) High-frequency ion source; 2) extracting electrode; 3) accelerating electrode; 4, 12) needle valves; 5) gas system; 6, 11) deflecting capacitor; 7, 16, 20) collector for measuring the beam current; 8, 10) vacuum valve; 9) tank of the plasma device; 13) stripping chamber; 14) capacitor for deflecting and modulating the ion beam; 15) electrostatic ion-energy analyzer; 17) window for adjusting the system; 18) chamber of the mass analyzer; 19) electromagnet; 21) ion-electron converter.

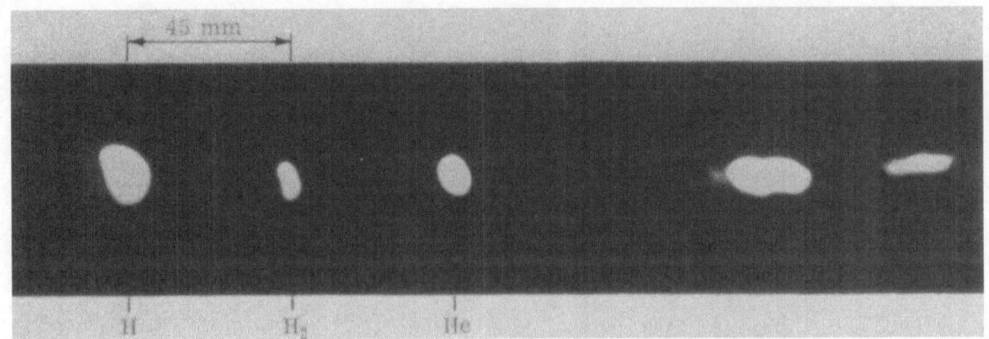

Figure 2. Pattern of the beam resolution by means of the magnetic mass analyzer.

the modulating voltage to the photocathode reduces the level of high-frequency noise (generated by the modulating voltage) by about one order of magnitude, compared to the usual circuits.

Generation of Multicomponent Neutral Beam

In order to generate a multicomponent beam of neutral particles comprising, H, H_2, and He atoms, a gas mixture of the composition 90% He + 10% H_2 was introduced into the source space through a needle valve.

Figure 2 shows the pattern obtained on a glass plate coated with a luminophor. This plate was placed at the point of the photomultiplier. The total beam consists of clearly resolved H^+, H_2^+, and He^+ beams and of almost unresolved beams of heavy ions (evidently, N^+, O^+, Si^+, and others).

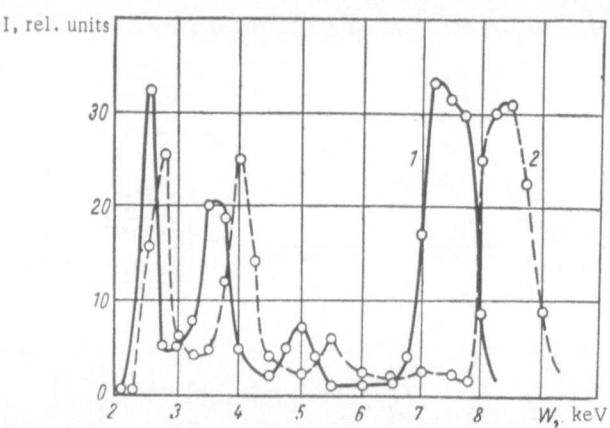

Figure 3. Energy spectrum of the hydrogen beam
when a voltage of (1) 7 kV and (2) 8.5 kV is applied
to the accelerating electrode.

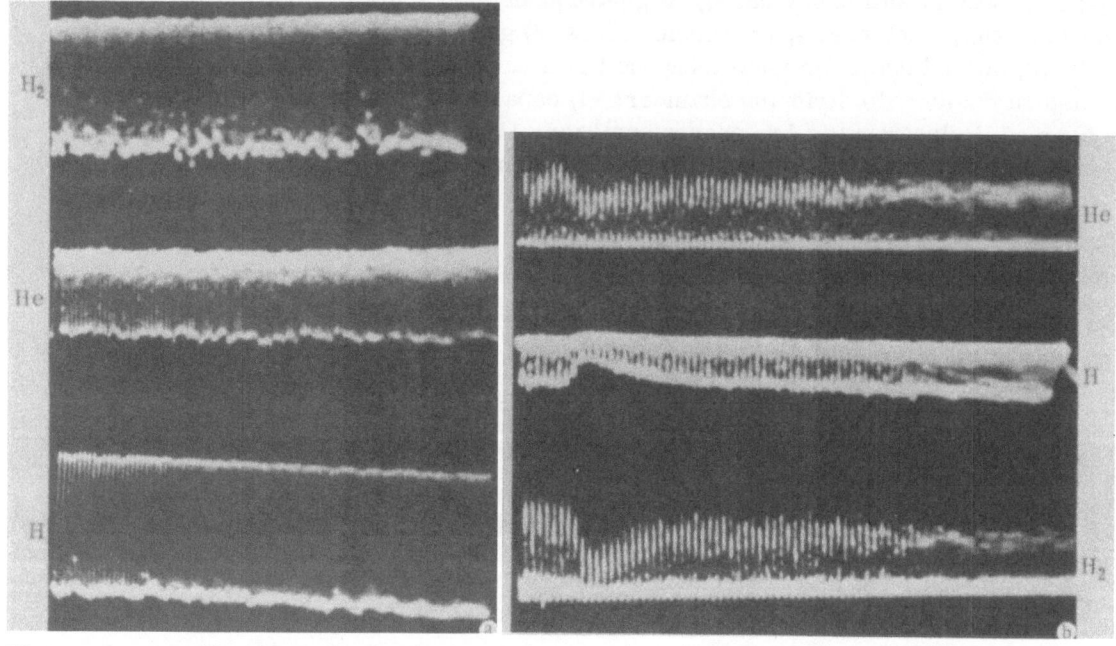

Figure 4. Oscillograms of signals resulting from H, H_2, and He beams (Figure a) and
oscillograms obtained when the beams are attenuated in a plasma burst (Figure b); energy of the beams, 7 keV.

The intensity ratio of the H^+, H_2^+, and He^+ beams at the output of the mass analyzer is approximated as follows:

$$J_{H^+} : J_{H_2^+} : J_{He^+} = 100 : 10 : 1.$$

The absolute value of the total beam intensity amounts to 10^{-7}-10^{-6} A.

The energy spectrum of the hydrogen beam recorded with the analyzer at a relatively
large energy width of the slits (±10%) is shown in Figure 3. The form of the energy spectrum can be easily explained by the collision-induced dissociation of fast neutral H_2^0 molecules
and H_2^+ and H_3^+ ions.

Figure 4 shows typical oscillograms of the signals generated by H^+, H_2^+, and He^+ beams modulated with the frequency 200 kHz. Additional experiments have shown that the beam-intensity fluctuations can be greatly reduced by proper selection of the operating conditions of the high-frequency discharge in the ion source.

Finally, Figure 4b shows typical oscillograms resulting from the attenuation of neutral He^0, H^0, and H_2^0 beams which are passed through a plasma burst generated by a conical-inductive source (pulsed admission of the gas [10]). The maximum ion density in the burst was about $2 \cdot 10^{14}$ cm^{-3}. The electron temperature in the maximum of the plasma-burst density was calculated from the attenuation of the helium beam and amounted to 70 ± 15 eV. The electron temperature calculated from the attenuation of the beam of fast hydrogen molecules was 60 ± 15 eV.

REFERENCES

1. O. V. Kozlov and V. D. Rusanov, Yadernyi Sintez, 4(4):312 (1964).
2. L. I. Krupnik et al., Zh. Tekh. Fiz., 35(4):711 (1965).
3. H. P. Eubank, Nuclear Fusion, 5(1):68 (1965).
4. V. V. Afrosimov et al., Zh. Tekh. Fiz., 36(1):89 (1966).
5. A. V. Chernetskii et al., "Apparatus and Methods for Plasma Research" [in Russian], Moscow, Atomizdat (1965).
6. G. R. Rik, "Mass Spectroscopy" [in Russian], Moscow, GITTL (1953).
7. A. K. Val'ter et al., "Electrostatic Accelerators of Charged Particles" [in Russian], Moscow, Gosatomizdat (1963).
8. Hughes and Rojansky, Phys. Rev., 34:284 (1929).
9. V. V. Matveev and A. D. Sokolov, "Photomultipliers in Scintillation Counters" [in Russian], Moscow, Gosatomizdat (1963).
10. N. I. Alinovskii et al., Zh. Tekh. Fiz., 36:877 (1966).

FLIGHT-TYPE MASS SPECTROMETERS
FOR RESEARCH ON PLASMA BURSTS

N. I. Alinovskii

The mass spectrometry of plasma bursts has to do with ions which are characterized by a large energy spread reaching from fractions of an electronvolt to several kiloelectronvolts. This fact is important for the design of mass spectrometers used for plasma burst studies. Employed are the mass spectrograph of Thomson [1, 2] and flight-type mass spectrometers [3, 4] which render additional information on the energy spectrum of the particles.

The present article describes two types of flight-type mass spectrometers which are used for research on plasma bursts in the Institute of Nuclear Physics of the Siberian Branch of the Academy of Sciences of the USSR.

1. A flight-type mass spectrometer similar to that described in [3] was used to investigate plasma bursts generated by a conical $\odot$ pinch [5].

The instrument consists of a flight tube (with a system extracting ions from the plasma and with a shutter) and a differential electrostatic energy analyzer for charged particles. Usually, a plane [3] or cylindrical [6] capacitor is used as the analyzing element. A narrow package of ions, separated by an electrostatic shutter from a long plasma burst, is scattered in its flight, depending upon the velocity distribution of the particles. The analyzer records only particles within a narrow energy interval. When the energy, the length of the flight path, and the time of flight of the ions are known, one can easily determine the mass-to-charge ratio of the ions.

Figure 1 shows the scheme of the experimental setup. A plasma burst, which is generated by a conical $\odot$ pinch source with pulsed admission of the gas (0.1 cm^3 at P = 500-700 mm Hg), propagates along a glass vacuum chamber with a diameter of 16 cm and a length of about 4 m. This chamber is placed on the axis of a system of magnetic bottles (T $\sim$ 5 msec, H = 0.5-2 kOe, plugging ratio $\alpha \sim 1.4$).

A differential electrostatic energy analyzer for charged particles, which is provided with a cylindrical capacitor (aperture angle 127°17') similar to that described in [6] is mounted opposite to the source in the vacuum chamber. This capacitor has an operating range extending from several electronvolts to several kiloelectronvolts. The distance between the tubes of the cylindrical capacitor is 2 cm. The radius of the average ion trajectory is 10 cm. The energy interval selected is given by the total width s of both the input and output slits:

$$\frac{\Delta E}{E} = \frac{s}{r},$$

where r denotes the average radius of the tubes.

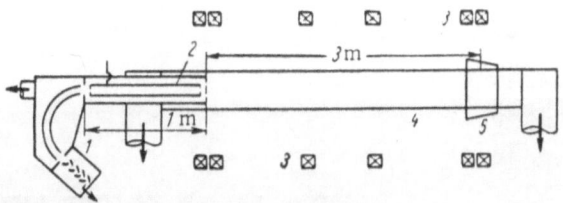

Figure 1. Scheme of the experimental device.
1) Electrostatic energy analyzer for the charged
particles; 2) extractor tube; 3) "Probkotron"
(magnetic bottle); 4) glass tube; 5) conical coil.

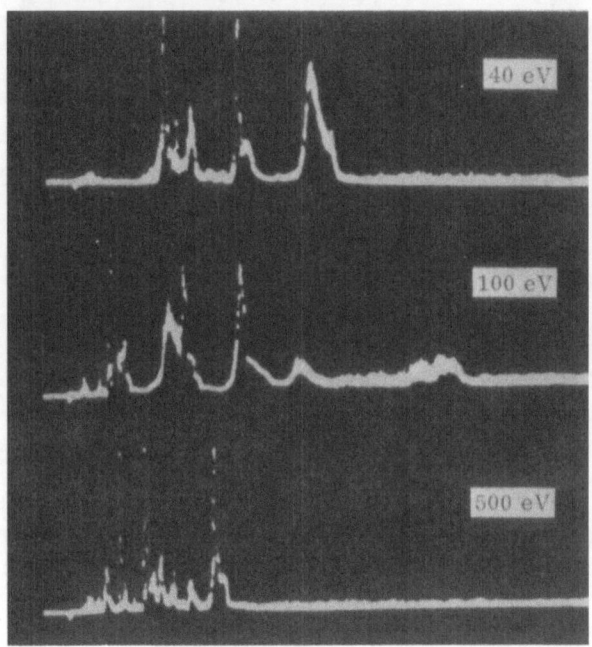

Figure 2. Typical current oscillograms of the
ionic plasma component for various particle energies. Time base, 350 μsec.

Figure 1 shows the flight tube for extracting charged particles from the plasma. An extracting potential of the order of 5 kV is applied to the internal part of tube 2. Thus, the ions
obtain an additional energy, which they lose again when they leave the energy analyzer, and pass
through the interior of the tube.

A VEU-OT-8M electron multiplier is used for recording the charged particles. When the
polarities of the extracting, analyzing, and multiplier-supplying voltages are properly chosen,
both ions and electrons can be recorded with this multiplier. Great difficulties with absolute
measurements are the principal disadvantage of this method. These difficulties arise because
the secondary electron emission coefficient of the first multiplier dynode is unknown for positive ion bombardment. Determinations of the relative concentrations of various ion components
are for the same reason difficult in any mass analysis.

The vacuum chamber of the device itself is used as the drift space. Between the point of
generation and the point of recording, the particles have to travel a path of about 4.5 m. The

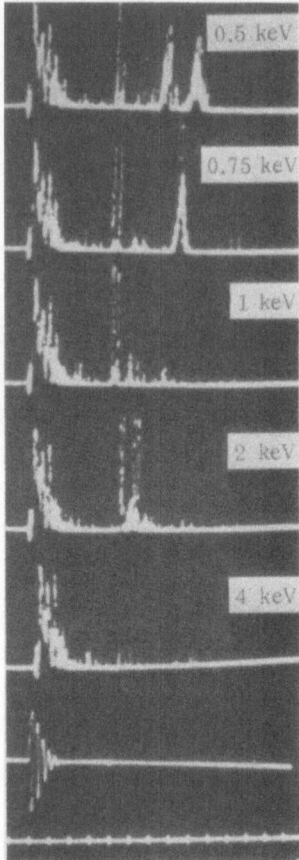

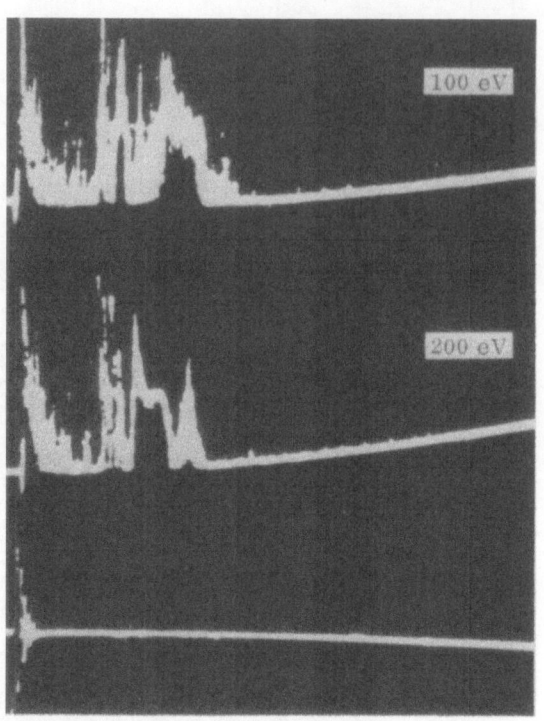

Figure 3. Typical oscillograms of the current corresponding to the electron component of the plasma; various electron energies. The lower trace corresponds to the source current. Time marks at 10-μsec intervals.

Figure 4. Typical oscillograms of the current originating from the electron component in the plasma. The trace at the bottom corresponds to the source current; time base, 350 μsec.

large transport length guarantees a rather high accuracy of both the spatial position of the particle source and its time-dependent parameters. Therefore, the system can be operated as a mass spectrometer.

Figure 2 shows typical oscillograms of the signals which originate in the electron multiplier from the ion component of the plasma. Three principal peaks, which approach the beginning of the discharge when the energy to be analyzed increases, are visible. When the ion energy is known and when the ion velocity has been determined from the flight time, one can assign these peaks to deuterium (D^+) ions, carbon (C^+) ions, and silicon (Si^+) ions. The ions of the C^+ and Si^+ admixtures originate during the discharge and are released from the surface of the glass chamber in the Θ pinch source. In this respect, the source becomes effective at the beginning of the third half-period of the discharge current [7].

Figures 3 and 4 show typical oscillograms of signals which were obtained with the electron multiplier when the electron component of the plasma was recorded. The first peak on all oscillograms lasts as long as the discharge current flows and results from the discharge radiation in the vacuum ultraviolet. A portion of this radiation is scattered and reflected at the external walls of the cylindrical capacitor, and hence, is incident on the output slit of the analyzer. This radiation knocks off electrons and is therefore recorded.

The energies of the electrons recorded seem to be related to the temperature, because the longitudinal electron energy, which, due to the quasi-neutrality condition, corresponds to the longitudinal ion velocity, is much smaller than the longitudinal ion velocities.

The results obtained with the electron analyzer make it possible to construct profiles of the electron and ion densities in the plasma burst.

Each oscillogram represents an elementary trapezium in the energy spectrum of the particles (with the spectrum as a function of time). The base of each trapezium is given by the energy width ΔE of the slit. Therefore, one can write

$$\int_0^\infty n v s\, dt = \int_{E-\frac{\Delta E}{2}}^{E+\frac{\Delta E}{2}} \frac{dN}{dE}\, dE \sim \overline{\frac{dN}{dE}}\, \Delta E,$$

where $E^\bullet$ denotes the energy at which the analysis is made.

Since $\Delta E \sim E$, we have

$$\frac{dN}{dE} \sim \frac{\int_0^\infty n v s\, dt}{E}.$$

Our calculations involve only relative quantities, and therefore, $\int_0^\infty n v s\, dt$ can be represented by the area covered by all peaks of an oscillogram. For calculating dN/dE, we normalize the area with respect to the energy by assuming the normalization factor to be equal to unity for some energy value, e.g., for $E = 1$ keV. In calculations referring to electrons, the relation between the efficiency of the electron count with the electron multiplier and the electron energy is taken into account. This relation is assumed to be roughly linearly dependent upon the incident electron energy and ranges from 100% at $E = 500$ eV (the voltage at the first multiplier dynode was equal to +500 V) to 40% at $E = 6$ keV [8].

Before constructing the profile of the density in the burst, the relation $nv(t)$ is graphically constructed. To do this, all oscillograms of $nv = \varphi(t)/s = f(t)$ are split into equal, sufficiently small time intervals Δt:

$$\Delta t_1 = t_2 - t_1; \quad \Delta t_2 = t_3 - t_2;$$

etc.

We determine for each t_i

$$\frac{dN}{dE_k} \sim \frac{\int_{t_i}^{t_i+\Delta t_i} n v s\, dt}{E_k}$$

and plot the curve

$$nv(t_i) \sim \sum_k \frac{dN}{dE_k}(t_i)\, \delta E_k,$$

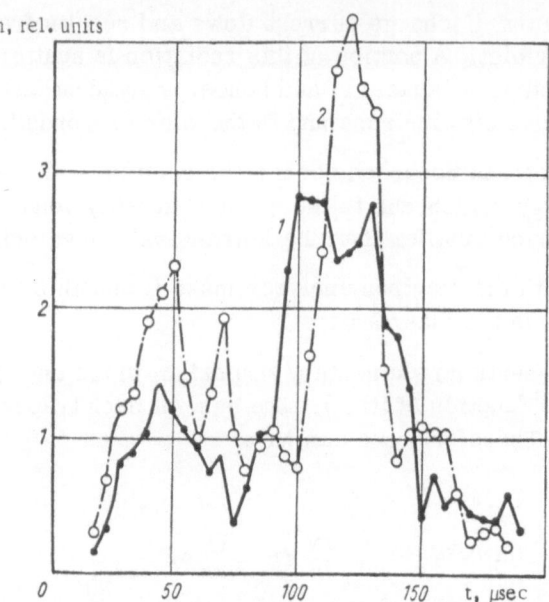

Figure 5. Density profile of the electron component in the plasma for (1) H = 200 Oe, and (2)
H = 430 Oe.

where $\delta E_k = E_{k+1} - E_k$; E_k denotes the energy values at which the measurements are made.

The profile of the electron and ion velocities as a function of time can be easily determined from the times at which the particles are recorded. The velocity can be represented in
the form of a hyperbola

$$v = \frac{l}{t}.$$

This relation is strictly correct for electrons and is an approximation for ions (due to the pres-
ence of the extracting tube).

When we divide graphically nv(t) by v(t), we obtain the density profile of the electrons
(Figure 5) and of the ion component of the plasma burst. Due to the quasi-neutrality condition,
the profiles agree. It follows from Figure 5 that the plasma burst consists of two particle
groups which move with the average velocities $5.5 \cdot 10^6$ and $2.2 \cdot 10^6$ cm/sec.

The profile of the electron density refers to the point at which the plasma burst enters
into the extracting tube, because before that point the electrons move with the longitudinal velocities of the ions (due to the quasi-neutrality condition).

The resulting data can be used to plot both the integral energy spectrum $\frac{dN}{dE}(E)$, and the
differential energy spectrum $\frac{dN}{dE}(E, t)$ of the particles in the plasma burst. The differential
energy spectrum of the ions can be used to construct a relation between the average direction-
dependent energy of the ion motion and the ion position in the burst (Figure 6). This relation
resembles that which would hold if the plasma burst would simply leave the source, i.e., we
have

$$E \sim \frac{1}{t^2}.$$

It is difficult to determine similar relations for the electrons because the fast primary
particles release a tremendous number of slow electrons with an energy spectrum ranging from
0 to 50 eV from the grids and the entrance diaphragms [9].

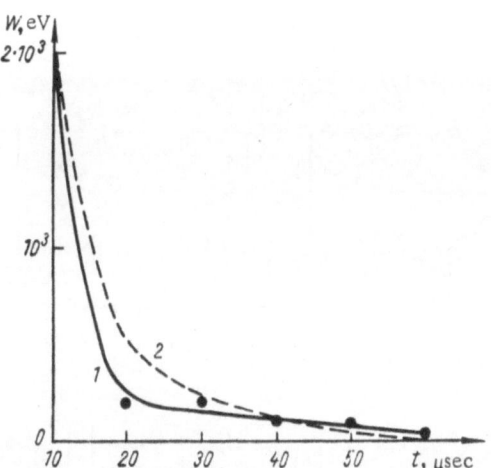

Figure 6. Profile of the direction-dependent average energy of the ions in the plasma burst (curve 1); the relation W ~ $1/t^2$ (curve 2).

The differential energy spectrum of the electrons indicate that the burst's front portion, which appears in the time interval between 20 and 40 μsec (time reckoned from the beginning of the discharge in the source) at the entrance of the extracting tube (see Figure 5), is characterized by an electron temperature of about 100 eV. This result agrees quite well with the temperature estimates made in [10] for a plasma burst obtained in the same fashion.

2. Figure 7 shows schematically another flight-type mass spectrometer. A vacuum system is provided at the entrance to the device; the system comprises a cone-shaped cap of stainless steel with a small-diameter opening (about 200 μ) at the apex.

Ions are extracted from the low-pressure plasma burst by means of an electric field and are accelerated. In the flight chamber, which has a length of about 2 m and to which an accelerating potential is applied, the ions move with the velocity

$$v_i = \sqrt{\frac{2q_i(U_{\text{acc}} + U_{\text{init}})}{M_i}},$$

where q_i and M_i denote the charge and the mass of the ion, respectively; q_iU_{init} denotes the initial ion energy; and U_{acc} denotes the accelerating potential.

An electrostatic capacitor-type shutter is mounted close to the accelerating system. This shutter forms ion packages with a length of 0.1 μsec. The shutter is switched on after an adjustable delay with respect to the start of the discharge in the source (delay to 0 to 1000 μsec).

The quantity M/q of the ions can be determined from the mass spectrograms corresponding to the "tail" sections of the burst, because for this burst section, the condition

$$qU_{\text{init}} \ll qU_{\text{acc}}$$

is readily satisfied. The initial ion energy can be determined when one considers the calculated flight time and sets $qU_{\text{init}} = 0$; then

$$E_{\text{init}} = qU_{\text{init}} = qU_{\text{acc}}\left(\frac{t_{\text{calc}}^2}{t_{\text{fl}}^2} - 1\right).$$

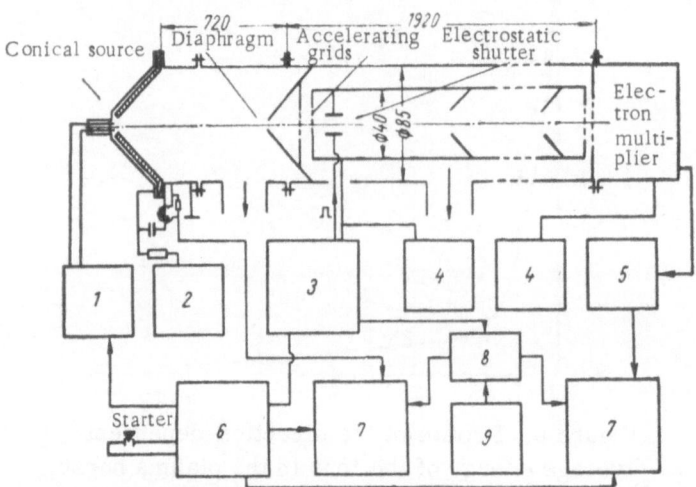

Figure 7. Scheme of the flight-type mass spectrometer.
1) Trigger generator; 2) rectifier, +10 kV; 3) block
for the deflecting voltage and circuit for generating
a rectangular shutter pulse; 4) rectifier, −5 kV; 5)
cathode follower; 6) GI-4M trigger circuit; 7) OK-17M
oscilloscope; 8) divider; 9) time-mark generator.

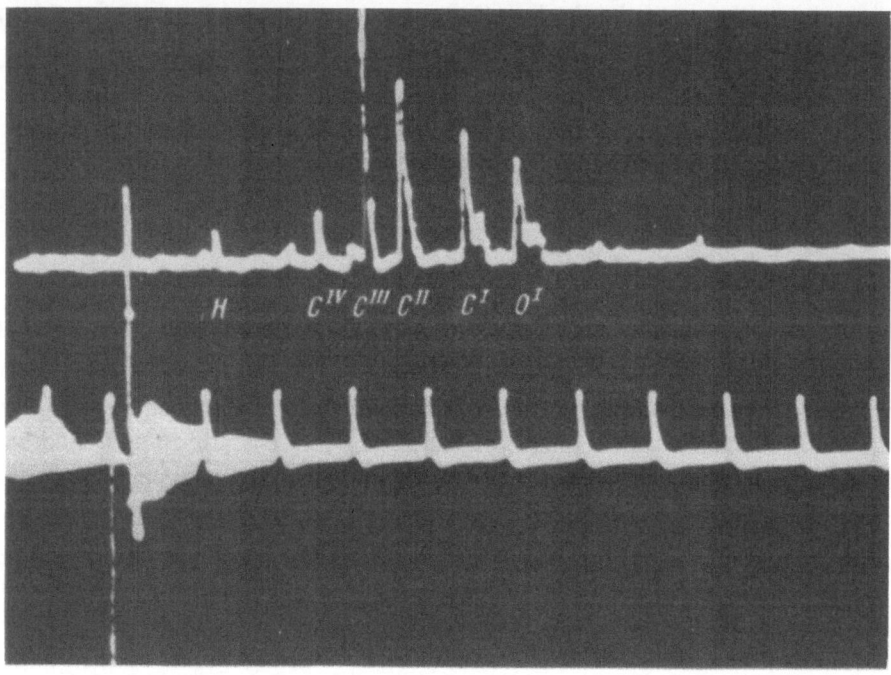

Figure 8. Typical mass spectrogram. The lower trace represents the
differentiated shutter pulse; time marks in 1.65-μsec intervals.

The relative error increases rapidly with decreasing energy, as is obviated by the formula

$$\frac{\Delta E_{\text{init}}}{E_{\text{init}}} = \frac{2t_{\text{calc}}^2 \Delta t}{t_{\text{fl}}\left(t_{\text{calc}}^2 - t_{\text{fl}}^2\right)},$$

where t_{calc} denotes the calculated flight time; t_{fl} is the actual flight time; and Δt is the error in the time determination.

Two mass spectrograms corresponding to a certain section of a plasma burst but to different accelerating potentials are required for a unique determination of M/q and qU_{init} of the ions. It is easy to show that

$$\frac{M}{q} = \frac{2t_1^2 t_2^2 (U_1 - U_2)}{l^2 \left(t_2^2 - t_1^2\right)}$$

and

$$E_{\text{init}} = qU_{\text{init}} = \frac{t_1^2 U_1 - t_2^2 U_2}{t_2^2 - t_1^2},$$

where t_1 and t_2 denote the flight time; U_1 and U_2 are the accelerating potentials; and l denotes length of the flight path.

The resolving power of this mass spectrometer is given by the ratio $M/\Delta M = t/2\Delta t$ and amounts to 20-40 in the range of medium masses.

In the case of these particular mass spectrometers, the ratio $\delta/\Delta t$ is very appropriate for describing the resolving power. The interval between signals corresponding to similar masses is denoted by δ and Δt denotes the observed signal duration. Obviously for $m_i \gg 1$, we have

$$\delta = \frac{t_i}{2m_i},$$

where t_i denotes the time of flight of an ion; and m_i is the ion mass expressed in units of the proton mass.

For medium masses, the relation $\delta/\Delta t > 1$ holds and characterizes the resolution in this mass range.

A VEU-OT-8M electron multiplier with dynodes of beryllium bronze is used for recording the ions as in the preceding mass spectrometer.

Figure 8 shows a typical mass spectrogram of a certain section of a plasma burst. The second oscilloscope trace shows the time marks and, in addition, the differentiated shutter pulse. The shutter pulse and the signal corresponding to the discharge current (derived from a Rogovskii flange) was applied to another oscilloscope in order to determine the position of the plasma burst portion analyzed.

An independent vacuum system with a titanium pump is provided for the mass spectrometer. The operating pressure in the chamber is maintained at $(5-7) \cdot 10^{-6}$ mm Hg.

This mass spectrometer was used for investigations of plasma bursts generated in a conical source in which a discharge took place along the surface of a plastic glass member [4]. The cone had an aperture of 25°. The parameters of the capacitor battery were C = 33 μF and L = 12 cm; the discharge current had an almost aperiodic time dependence.

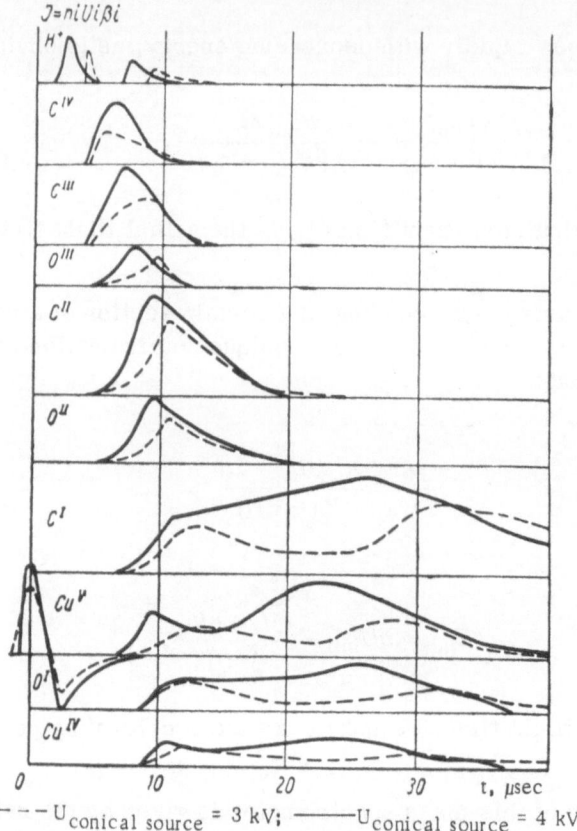

Figure 9. Mass spectrogram of a plasma burst
at the following source voltages: ———) 4 kV and
– – –) 3 kV.

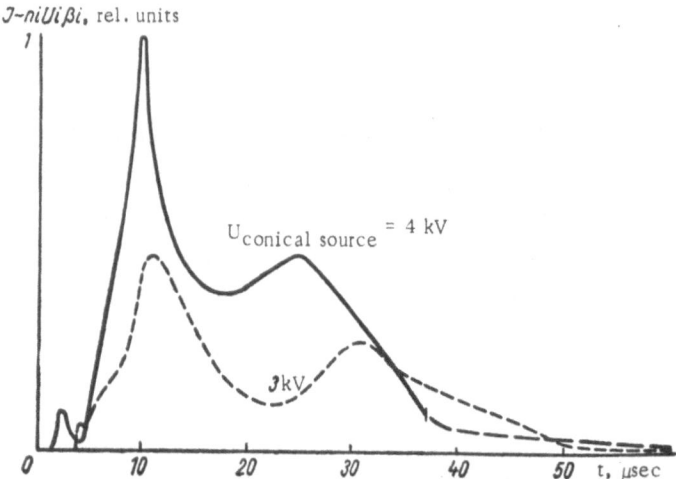

Figure 10. Profile of the current corresponding to
the plasma ions.

This article describes the principal experimental results which were obtained with two different voltages applied to the capacitor battery (3 and 4 kV) and with negative initial polarity of the inverse current line.

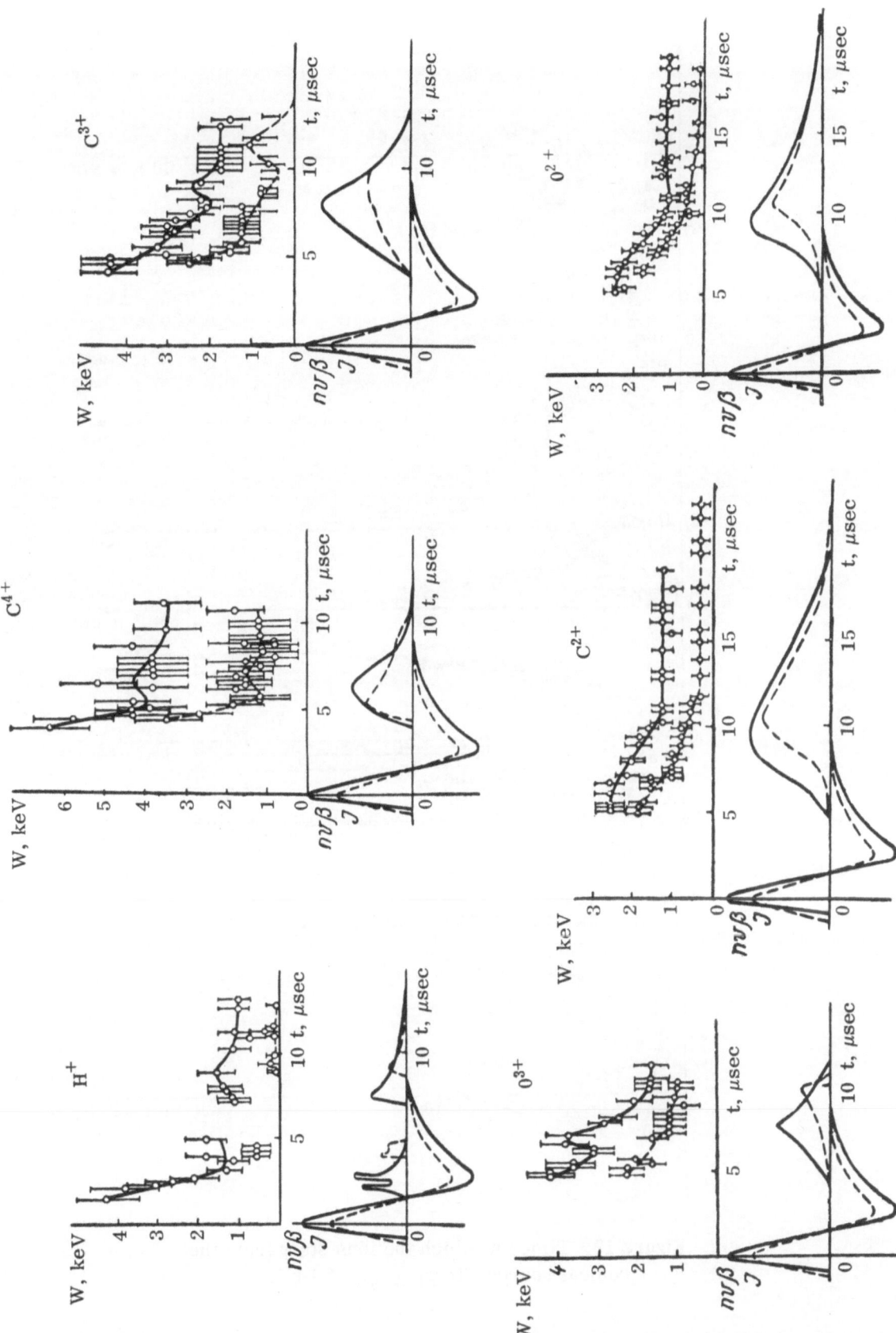

Figure 11. Energy distribution of the ions over the burst at the following source voltages: (———) 4 kV and (— — —) 3 kV.

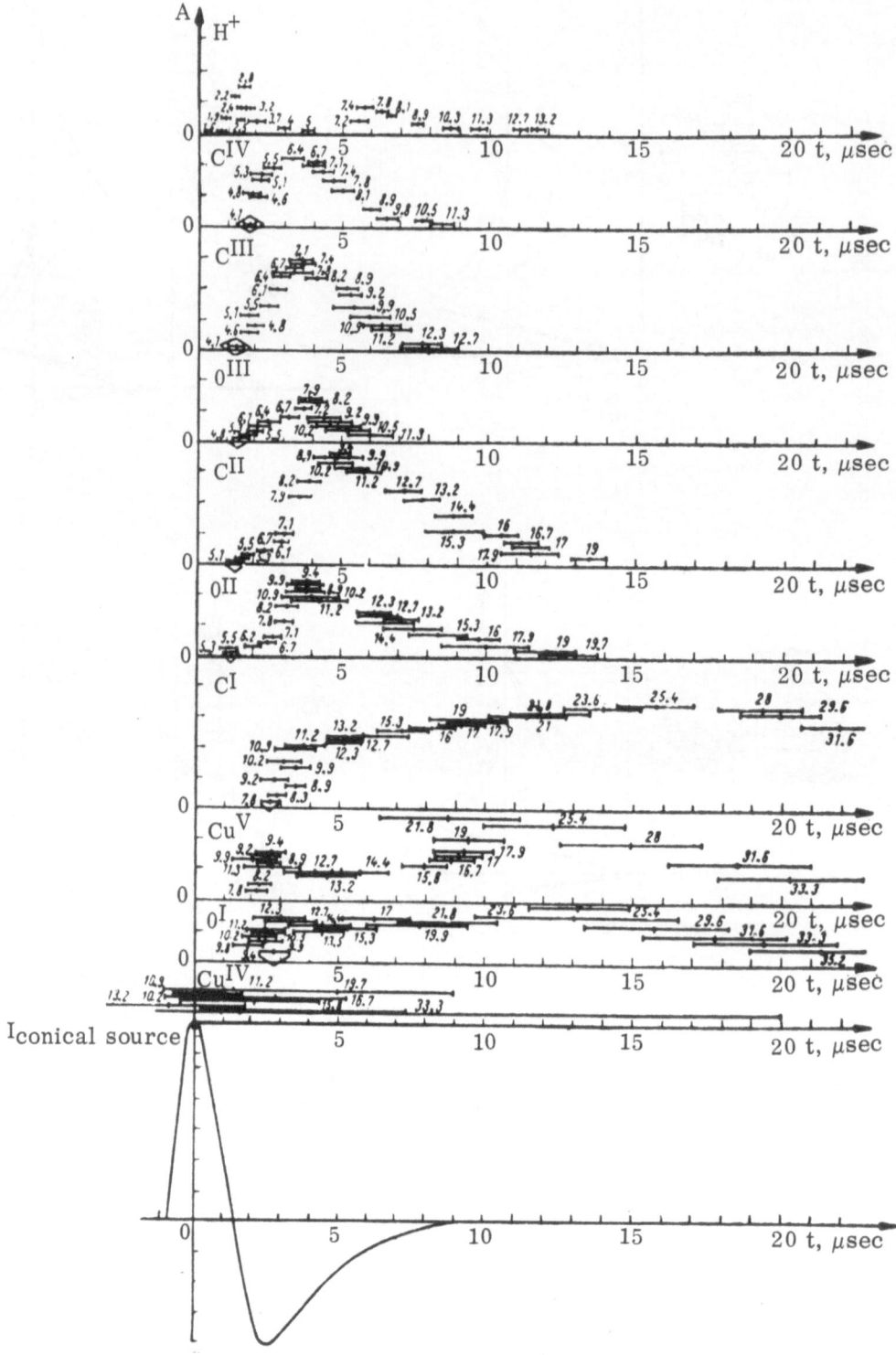

Figure 12. Times at which the ions start from the
conical source. $U_{conical\ source} = 4$ kV.

The distance between the conical source and the mass spectrometer was 72 ± 2 cm in all experiments. The conical source was evacuated by means of a titanium pump which kept the pressure in the system at the value $(1-3) \cdot 10^{-6}$ mm Hg.

Figure 9 shows the total mass spectrogram of a plasma burst; the fine layer structure of the burst is clearly visible.

By superimposing the signal amplitudes at certain times on this mass spectrogram, the overall burst profile shown in Figure 10 is obtained.

It follows from this figure that at a distance of 72 cm from the source, the plasma burst is split into three groups, namely an initial burst (H^+), a secondary burst (H^+, C^{4+}, C^{3+}, O^{3+}, C^{2+}, and O^{2+}), and a third burst (C^+, Cu^{5+}, O^+, and Cu^{4+}). A similar pattern was obtained by several researchers [3] working with other sources.

Figure 11 shows the energy distribution of the ions over the burst and the ion profiles which are proportional to $n_i v_i \beta_i$, where n_i denotes the ion density, v_i the ion velocity, and β_i the coefficient of secondary electron emission from the first multiplier dynode when the dynode is bombarded by positive ions. The error resulting from the finite length of the shutter pulse and the signal at the mass spectrometer is indicated at each experimental point. The energy distribution pattern is typical for all ions. The pattern is characterized by a rather sharp decline from the high-energy region and by a transition to a plateau value.

Once the ion velocity has been determined from the ion-energy distribution, and the flight-path length and the ion-arrival times (reckoned from some characteristic point of the discharge current curve) are known, one can determine the time at which the ions start from the source. The following assumptions are made: (1) the ions move with a constant velocity in the conducting plasma; and (2) the acceleration occurs during a short time interval which is small compared to the time of the discharge, and the acceleration takes place within or very close to the source.

Figure 12 shows the results of a corresponding calculation. The error considerations take into account the inaccuracy in the determination of the ion energy as well as the undetermined starting point of the particles in the source.

The time at which a particular type of ions arrives at the mass spectrometer is indicated around each point (time expressed in microseconds). The time was reckoned from the first maximum of the discharge current curve.

A detailed analysis of the resulting experimental data has been given in [4]. There, several general conclusions and assumptions concerning the acceleration of plasma bursts have been made.

REFERENCES

1. S. Yu. Luk'yanov et al., Zh. Tekh. Fiz., 33:1926 (1961).
2. I. I. Konovalov, L. I. Krupnik, et al., in: "Plasma Diagnostics" [in Russian], Moscow, Gosatomizdat (1963).
3. A. A. Kalmykov et al., in: "Plasma Diagnostics" [in Russian], Moscow, Gosatomizdat (1963).
4. N. I. Alinovskii and Yu. E. Nesterikhin, Report of the Conference on Plasma Injectors [in Russian], Khar'kov (1964).
5. A. M. Iskol'dskii et al., Zh. Eksper. i Teoret. Fiz., 47(8):771 (1964).
6. H. P. Eubank and T. D. Wilkerson, Rev. Sci. Instr., 34(1):12 (1963).
7. A. M. Iskol'dskii et al., Prikl. Mekhan. i Tekh. Fiz., No. 6, p. 119 (1965).
8. A. I. Akishin, "Ion Bombardment in Vacuum" [in Russian], Moscow, Gosenergoizdat (1963).

9.   L. N. Dobretsov, "Electron and Ion Emission" [in Russian], Moscow, Gostekhizdat (1952).
10.   N. I. Alinovskii et al., Zh. Tekh. Fiz., 36:877 (1966).

EXPERIMENTAL METHOD FOR INVESTIGATING THE
ENERGY LOSSES AND THE STABILITY OF A PLASMA
IN TOROIDAL CHAMBERS OF THE TOKAMAK TYPE
BY MEANS OF A BOLOMETRIC PROBE

L. L. Gorelik and V. V. Sinitsyn

It is well known that when a plasma undergoes ohmic heating in toroidal chambers to which a strong magnetic field has been applied, a strong plasma outflow occurs even in the case of a rather large hydromagnetic stability reserve [1-3]. This outflow can be described by the Kruskal–Shafranov criterion, which states that the so-called macroscopical plasma stability is reached when no oscillations occur in the overall characteristics of the discharge. However, until recently, no adequate methods for evaluating the principal energy losses of plasmas in toroidal devices were available. Furthermore, since the lifetime of the charged particles in plasmas are not clearly defined (the densities can be determined with radio-interferometric measurements of the density τ_p), one cannot establish a relation between the lifetime and the plasma parameters. The present article relates to the TM-2 device [4] and describes: (1) a method for investigating the principal energy losses and the energy balance in a plasma by means of a bolometric probe, and (2) an experimental method, based on bolometric determinations of the energy losses in a plasma, for determining the lifetimes τ_E of the charged plasma particles.

Method for Investigating the Energy Losses

in a Plasma

The investigation of the energy losses in a plasma by means of a bolometric probe resembles, in principle, the technique described in [5]. The design of the probe is shown in Figure 1. A three-layer germanium bolometer is used as the detector [6]. The detector is attached to the front side of a one-meter-long, stainless steel tube with an external diameter of 8 mm. The tube is attached to the base of a socket provided for measurements on the chamber. The seal is so that the bolometer can be placed at various distances from the surface of the plasma string without need for disturbing the vacuum. The tube of the probe serves also as the external conductor of a coaxial cable over which the bolometer signal is fed to a cathode follower. The signal derived from the cathode follower output is applied to the input of an OK-24 oscilloscope. The center conductor of the coaxial cable is kept in fixed position by means of ten Teflon disks placed at equidistant intervals over the entire length of the tube. This design helps to avoid errors created by mechanical shocks of the probe. An opening for the pump is provided in the tube so that the tube interior can communicate with the vacuum chamber of the device. The design of the probe prevents charged particles and radiation from being incident

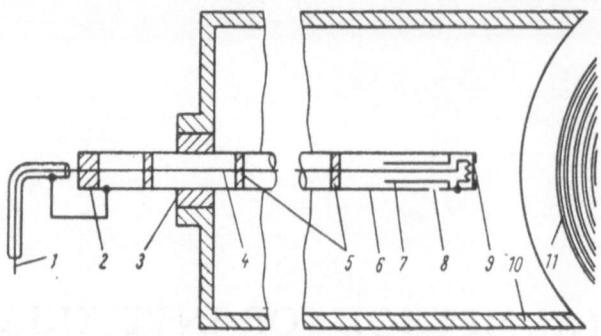

Figure 1. Design of the probe and its position in the socket for measurements. 1) Output to the cathode follower; 2) insulating seal; 3) flexible Teflon seal; 4) center conductor; 5) Teflon disk; 6) tube of the probe; 7) internal shield; 8) opening for evacuation; 9) germanium bolometer; 10) attachment for measurements; 11) plasma string.

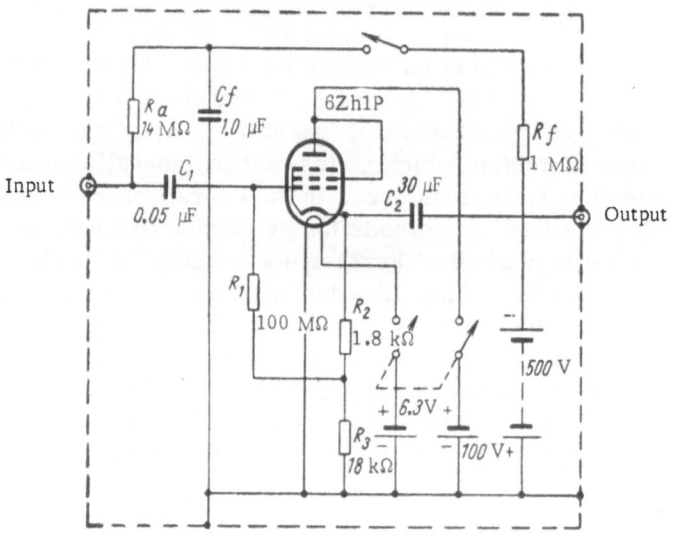

Figure 2. Cathode follower circuit.

on the probe. The cathode follower (Figure 2) comprises a 6Zh1P finger pentode. In order to avoid inductive disturbances generated by the discharge current, the cathode follower receives its supply voltages from dry cells. The bolometer supply circuit, consisting of a 500-V battery, an additional R_a, and a decoupling filter R_f, C_f, is housed in a single block also containing the cathode follower. The input resistance of the cathode follower with the connected bolometer supply circuit amounts to 7 MΩ; the output resistance of the cathode follower is 2 kΩ. With these conditions of operation, the cathode follower matches the resistance of the germanium bolometer (about 3 MΩ) to the input resistance of the OK-24 oscilloscope (2 kΩ).

The above-described measures guarantee a rather noise-free bolometric probe. The sensitivity threshold of the probe amounts to about 10^{-5} J/cm^2 and its resolving time about 300 μsec.

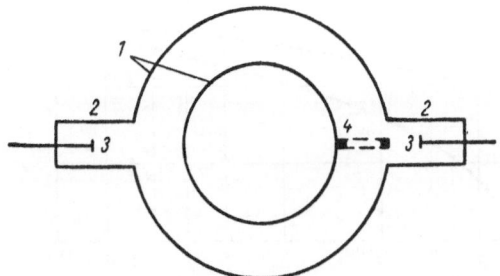

Figure 3. Position of bolometric probes
in the TM-2 device. 1) Linear; 2) chamber sockets; 3) bolometer probe;   4)
diaphragm.

The energy which is liberated per unit surface of the liner is given by the expression

$$E_k = K(h)\,E_b(h),$$

where h denotes the distance between the liner surface and the bolometer; E_0 is the calculated
energy which is obtained by dividing the energy supplied to the discharge by the liner surface;
E_b denotes the energy liberated per unit surface of the bolometer; and K(h) denotes a coefficient which takes into account the solid angle in which the plasma is visible from the bolometer.
The ratio E_k/E_0 corresponding to the end of the discharge was determined in the majority of
experiments. The relative error in the determination of this quantity was basically caused by
a systematic error and amounted to 10-15%. The measurements were made in two sections
of the discharge chamber, namely close to the diaphragm and far away from it (Figure 3).

A germanium bolometer with a nonblackened aluminum surface was used in the initial
stage of the investigations. Later, a bolometer with a surface blackened by evaporating bismuth in a hydrogen atmosphere under a pressure of 0.25 mm Hg was used. In order to estimate
the absorption of these bolometers, they were compared with another bolometer which had been
blackened by evaporating aluminum in a hydrogen atmosphere under a pressure of 2 mm Hg.
A bolometer of the latter type absorbs almost completely all radiation in a wide spectral range
extending to the far infrared.* It turned out that the ratio of the absorptive capacities of the
bolometers blackened with bismuth and blackened with aluminum is 0.6 for visible light and
close to unity for radiation emitted by an IFK-120 flash lamp. Furthermore, the absorptive
capacities of both the blackened and the nonblackened bolometer differed by less than 5% for
the radiation emitted by the flash lamp. This result gives reason to believe that nonblackened
bolometers can be used for measurements of the plasma radiation. As a matter of fact, the results of measurements which we made on the TM-2 device with nonblackened and bismuth-
blackened bolometers agreed rather strongly.

In order to obtain an idea of the principal energy losses, the losses were measured in a
cross section far from the diaphragm under various degrees of hydromagnetic stability and
for various h values. The results of these measurements indicate that in stable operation, the
charged particles do not leave the liner. In unstable operation, the particles leave the liner
in large numbers, but are not incident on the bolometer when this is withdrawn into the rear
part of the opening for diagnostic purposes. Thus, a bolometer, which has been removed from

* At pressures below 1 mm Hg, this bolometer has a high time constant when it assumes a uniform temperature over its entire thickness and was therefore not used for measurements on
   our device.

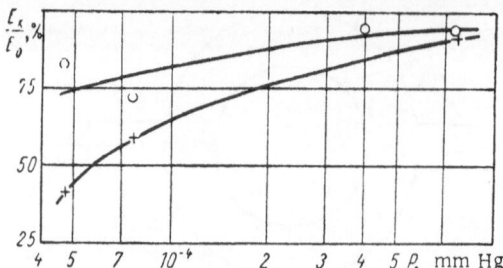

Figure 4. Dependence of E_k/E_0 upon the
initial argon pressure: + far away from the
diaphragm; ○ in the region of the diaphragm.

the surface of the plasma string, records the energy losses which result only from radiation
and neutral particles.

At high initial hydrogen pressures in the chamber (about 10^{-3} mm Hg) and rather large h
values (h ⩾ 30 cm), the conditions for operation of a "gas filter" are satisfied [5]. One can
assume that the bolometer does not record neutral particles with energies of 50-100 eV in this
case. Therefore, a comparison of E_k/E_0 values at h values close to zero with the corresponding
values at h values in excess of 30 cm permits one to estimate the contribution of the neutral particles to the total energy loss.

In these investigations of the energy losses in the plasma near the diaphragm, the influence of the solid angle cannot be estimated due to the shielding effect of the diaphragm at
short distances between the bolometer and the diaphragm. Thus, the bolometer probe had to
be calibrated. Bolometric measurements of the losses in the argon discharge were made in
order to evaluate the possibilities of a calibration. The dependence of E_k/E_0 upon the initial
argon pressure is shown in Figure 4. According to Figure 4, at argon pressures of about 10^{-3}
mm Hg, E_k/E_0 is close to 100% in both chamber cross sections. One can therefore assume
that in the case under consideration, the energy is uniformly incident on the chamber walls.
This makes it possible to perform a calibration. With this calibration, E_k/E_0 measurements
near the diaphragm can be used to obtain, for various h values, the loss distribution over the
liner surface near the diaphragm. For example, a localized source of plasma losses would
correspond to a decrease in E_k/E_0 with decreasing h. Results of measurements made on the
TM-2 device indicate that an area of additional losses exists near the diaphragm. It is possible to estimate the contribution of this area to the total energy losses.

The above-described method helped to obtain data concerning the principle energy-loss
components, their distribution over the surface of the discharge chamber, and the energy balance
in the TM-2 device.

Experimental Method for Determining the

Energetic Lifetimes of Charged Particles

As has been mentioned above, when the bolometer is far from the surface of the plasma
string, the energies carried away from the plasma by charged particles are not recorded. One
can therefore evaluate the energy losses E_{part}, created by charged particles, from the energy
supplied to the discharge and the radiation-induced and neutral particle-induced energy losses
measured with the bolometer. Furthermore, one can evaluate the quality of the thermal insula-
tion and the plasma stability from the electron temperature and the quantity E_{part}. In order to
obtain definite stability characteristics of the plasma and to compare experimental results with

theoretical values, it is useful to determine the lifetime of the charged particles from considerations of the energy balance in the plasma [1, 7]. The energetic lifetime, averaged over the discharge time $\bar{\tau}_E$, can be calculated with the formula

$$\bar{\tau}_E = \frac{\overline{N}\,[\gamma k\,(T_e + T_i) + eV_i]}{\varepsilon_{\text{part}}}\,t_p,\tag{1}$$

where T_e and T_i denote the electron and ion temperature at the current maximum, respectively (under the assumption that these values do not greatly differ from the values averaged over the duration of the process under consideration); γ is the ratio of the average energy transferred to the wall by an electron-ion pair to the temperature $(T_e + T_i)/2$; eV denotes the energy spent for ionizing a hydrogen atom; $\varepsilon_{\text{part}}$ denotes the energy losses resulting from charged particles during the discharge time (in the entire volume of the discharge chamber); $\overline{N}$ is the number of plasma electrons averaged over the discharge time; and t_p denotes the duration of the discharge. As has been shown by K. A. Razumov, at the time corresponding to the maximum of the discharge current, the energetic lifetime τ_E in the TM-2 device can be determined with the equation

$$\tau_E = \frac{N\,[\gamma k\,(T_e + T_i) + eV_i]}{I_p U - \dfrac{dQ}{dt} + eV_i\left(\dfrac{dN}{dt}\right)_{\text{pl}}},\tag{2}$$

where N denotes the total number or electrons in the plasma; I_p is the current through the plasma; U denotes the circumferential voltage; Q are the energy losses by radiation and neutral particles; and $(dN/dt)_{\text{pl}}$ denotes the rate of change of the number of electrons in the plasma. Under the experimental conditions used in the TM-2 device, the term $eV_i(dN/dt)_{\text{pl}}$ is much smaller than the difference between the first two terms in the denominator and hence, the term $eV_i(dN/dt)_{\text{pl}}$ can be ignored.*

When τ_E and $\bar{\tau}_E$ were determined with the above formulas, T_e could be found from the electric conductivity of the plasma, N from the electron density n_e measured with the radio-interferometric method and from the volume filled by the plasma, and dQ/dt from differentiation of the time-dependent bolometer curve for the energy losses along the liner surface. It was assumed for the calculations, that $T_i \approx T_e$. With this assumption, we set $\gamma = 3.5$. All quantities entering into Eqs. (1) and (2) were determined without taking into account changes in the inductance of the plasma string. A uniform plasma distribution and a uniform current distribution over the cross section of the plasma string were assumed in the determinations of τ_E and $\bar{\tau}_E$. Apart from this, it was assumed that both the plasma string and the current string have a radius equal to the internal diaphragm radius. Inspection of the above formulas reveals that even when in certain cases this assumption is not quite fulfilled, no great errors are made in τ_E and $\bar{\tau}_E$ determinations according to the above measurement method. By expressing the quantities T_e and N of Eqs. (1) and (2) by the radius a of the plasma string, one can show that these expressions depend upon a in the form $a^{-1/3}$. One can expect that the errors will not be very large in cases in which the region of current flow is roughly identical with the region in which the major part of the plasma is localized, and when the average radii of these regions (which are given by the distribution of both the plasma and the current in any cross section) are equal. We note that the requirement that the average radii of these regions be equal can be replaced with the less stringent requirement of a constant ratio of the radii, when relative estimates are made.

---

*When the quantity $\bar{\tau}_E$ is determined, certain assumptions which were made in deriving the equation for τ_E at the time of the current maximum need no longer be made.

The method outlined above was used to gather information on the energetic lifetimes of charged particles for various discharge parameters of the TM-2 device. Absolute calculations of the lifetimes could be made in these experiments with an accuracy equal to a factor of 2, whereas the accuracy in measurements of the relative changes amounted to 10 to 20% under various conditions.

The authors are indebted to L. A. Artsimovich and I. K. Kikoin for their interest in this work and for valuable remarks; to K. A. Razumova for valuable advice and assistance in the work; to B. B. Kadomtsev and V. S. Mukhovatov for helpful discussions; to E. P. Gorbunov for providing data on the electron density; to V. I. Nikolaev for his assistance in producing the equipment; and to L. S. Efremov and A. A. Kondrat'ev for their assistance in the measurements on the TM-2 device.

REFERENCES

1. E. P. Gorbunov and K. A. Razumova, Atomnaya Energiya, 15:363 (1963).
2. W. Stodiek, R. A. Ellis, and I. G. Gorman, Conf. on Plasma Physics and Controlled Nuclear Fusion Research, Salzburg, Report No. 31 (1961).
3. R. W. Mothley, Conf. on Plasma Physics and Controlled Nuclear Fusion Research, Salzburg, Report No. 167 (1961).
4. L. L. Gorelik, K. A. Razumova, and V. V. Sinitsyn, Report Given during the Conference on Plasma Physics, Culham, 1965.
5. L. L. Gorelik and V. V. Sinitsyn, Zh. Tekh. Fiz., 32:1406 (1962).
6. L. L. Gorelik and V. V. Sinitsyn, Zh. Tekh. Fiz., 34:505 (1964).

THERMOCOUPLE METHOD FOR DETERMINING THE ENERGY LOSSES IN A PLASMA AND THE DISTRIBUTION OF THE ENERGY LOSSES OVER THE DIAPHRAGM OF THE DISCHARGE CHAMBER

L. L. Gorelik and V. V. Sinitsyn

Bolometric measurements have shown that between 10% and 80% of the energy losses in a plasma (depending upon the conditions) occur directly on the diaphragm of the Tokamak TM-2 device [1]. In order to determine the magnitude of the losses and their distribution, we suggested using the diaphragm as a calorimeter and to record (with thermocouples) the temperature changes which develop after a discharge on various sections of the diaphragm. The distribution of the losses over the perimeter of the diaphragm can be established from the indications of thermocouples placed on the perimeter. The indications are used at a time t_f at which the temperature could level out over the width of the diaphragm, but could not yet reach the equilibrium value between neighboring thermocouples (Figure 1). In order to prevent the thermocouples from being broken, they can be attached far from the inner edge of the diaphragm; for example, at a medium perimeter line. An estimate shows that under normal conditions, the heating of the usual diaphragm by an energy amounting to 20-50% of the energy introduced to the system changes the diaphragm temperature by about 1°. Such changes can be easily detected with a sensitive microvoltmeter (e.g., an M-95 instrument with a scale division of $1 \mu V$) when one compares the readings before and after a discharge.

The time constant τ_b for attaining temperature equilibrium between two points of the diaphragm can be estimated with a formula of [2]:

$$\tau_b \approx \frac{c\rho}{10k} l^2, \tag{1}$$

where c, ρ, and k denote the specific heat, the density, and the heat conductivity of the diaphragm material, respectively; and l is the distance between the points on the diaphragm. Calculations made with this formula indicate that in the case of a diaphragm width $l \sim 2$ cm, τ_b amounts to about 0.7 sec for tungsten, and to about 0.8 sec for molybdenum. Taking into account that a rather extensive levelling of the temperature over the width of the diaphragm is necessary for obtaining high accuracy in absolute measurement and that, in addition, a certain time must elapse in order to take accurate readings on the output instrument (this time interval depends mainly upon the time constant of the output instrument), a time interval $t_f \sim 10$ sec must be tentatively assumed between readings. In the case of a tungsten diaphragm, the temperature levels out over a distance of about 5 cm during the 10-sec time interval. This means that the distance between the thermocouples on the perimeter L must be of the order of 10 cm. Our method permits one to determine both the loss distribution on the perimeter and the absolute loss figures from averaged indications of all thermocouples.

49

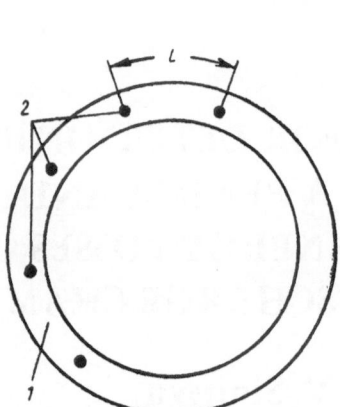

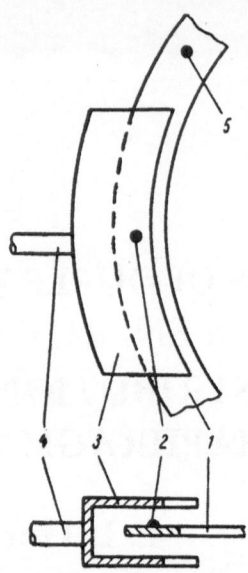

Figure 1. Locations of thermocouples 2 on the perimeter of diaphragm 1.

Figure 2. Design and position of the shielding screen. 1) Diaphragm; 2, 5) thermocouples; 3) screen; 4) handle for movement of the screen.

The loss distribution over the width of the diaphragm at some particular part of the diaphragm can be determined with a thermocouple placed on the center of the particular part and with a double shielding screen that can be moved in radial direction to cover a variable plasma section on both sides (Figure 2).* The screen must be thermally isolated from the diaphragm. The screen can be moved with a loosely attached handle. The dimensions of the shielding screen must be selected so that noticeable errors in the determinations of the loss distribution are avoided (these errors result from the temperature levelling between neighboring sections of the diaphragm perimeter). In the case of a tungsten diaphragm, the size of the screen covering the perimeter should be of the order of 10 cm.

The proposed thermocouple method was experimentally tested in work done in collaboration with N. D. Vinogradova and K. A. Razumova. A short description of the results of those experiments has been given in [1].

The authors express their gratitude to K. A. Razumova for valuable discussions.

REFERENCES

1.   L. L. Gorelik, K. A. Razumova, and V. V. Sinitsyn, Report presented at the Conference on Plasma Physics, Culham, 1965.
2.   L. L. Gorelik, Zh. Tekh. Fiz., 34:496 (1964).

---

*It appears possible to use for this purpose a special, movable diaphragm section with a thermocouple attached to it; this version is simpler from the design viewpoint.

INVESTIGATION OF THE SPACE- AND TIME-DEPENDENT EMISSION CHARACTERISTICS OF A PULSED TOROIDAL PENNING DISCHARGE

A. B. Berezin, D. G. Bulyginskii, and M. I. Vil'dzhunas

The present article is a continuation of studies dealing with the properties of pulsed toroidal Penning discharges [1]. Two discharge conditions have been previously established, i.e., a low-current condition and a high-current condition in which the current is limited by the resistance of the external circuit (under the experimental conditions, $I \simeq 200$ A). Strong oscillations were in both cases observed on the oscillograms of 8 and 4 mm UHF signals and of the hydrogen-ion emission. The present article outlines the results of a study of both the space- and time-dependent radiation characteristics and the electron temperature in the discharge.

The electron temperature T_e of the discharge was determined by comparing the relative intensity of the lines of the Balmer series in the discharge spectrum with the relative intensity of the same lines in the spectrum of a glow discharge. The electron temperature of the glow discharge was determined by the double probe method. The parameters of the glow discharge used for calibration were: hydrogen pressure, 10^{-1} mm Hg; voltage, 2.2 kV; and current, 20–40 mA. The distance between the electrodes of the double probe was 2 mm. The temperature calculated from the probe characteristics was 5 eV for a discharge current of 40 mA. The discharge parameters could be easily reproduced during the experiments. Under the assumption that the electrons of both the glow discharge and the toroidal Penning discharge follow a Maxwell distribution, one can derive the following formula by using Frish's approximation [2] for the excitation function of the H_α and H_β lines:

$$\left(\frac{I_\alpha}{I_\beta}\right)_{PD} \bigg/ \left(\frac{I_\alpha}{I_\beta}\right)_{GD} = \frac{\dfrac{\dfrac{1}{\sqrt{\tau_\alpha\,PD}} + \dfrac{3}{2}\dfrac{q_\infty^\alpha}{q_{max}^\alpha}\sqrt{\tau_\alpha}\,PD}{\dfrac{1}{\sqrt{\tau_\beta\,PD}} + \dfrac{3}{2}\dfrac{q_\infty^\beta}{q_{max}^\beta}\sqrt{\tau_\beta}\,PD}}{\dfrac{\dfrac{1}{\sqrt{\tau_\alpha\,GD}} + \dfrac{3}{2}\dfrac{q_\infty^\alpha}{q_{max}^\alpha}\sqrt{\tau_\alpha}\,GD}{\dfrac{1}{\sqrt{\tau_\beta\,GD}} + \dfrac{3}{2}\dfrac{q_\infty^\beta}{q_{max}^\beta}\sqrt{\tau_\beta}\,GD}},$$

where $(I_\alpha/I_\beta)_{PD}$ denotes the ratio of the H_α and H_β line intensities in the Penning discharge $(I_\alpha/I_\beta)_{GD}$ denotes the ratio of the H_α and H_β line intensities in the glow discharge; $\tau_{\alpha\,PD}$

51

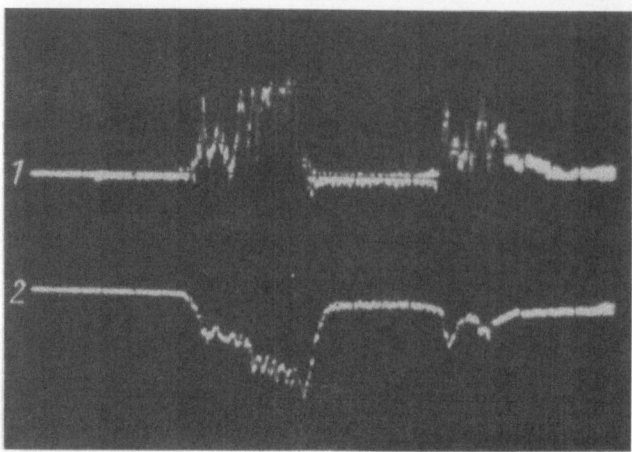

Figure 1. Correlation between (1) the probing 8-mm
UHF signal and (2) the light signal.

and $\tau_{\alpha GD}$ denote the ratio of the temperatures of the Penning discharge and glow discharge to the excitation energy of the upper level of the H_α line, respectively; $\tau_{\beta PD}$ and $\tau_{\beta GD}$ denote the ratio of the temperatures of the Penning discharge and the glow discharge to the excitation energy of the upper level of the H_β line, respectively; $q_\infty^\alpha/q_{max}^\alpha$, $q_\infty^\beta/q_{max}^\beta$ denote the ratio of the excitation functions of the H_α and H_β line at infinity to the values in their maximum, respectively.

The temperature of the Penning discharge was calculated with this formula for the discharge conditions: $P = 1 \cdot 10^{-3}$ mm Hg, $H_z = 3$ kOe, and $U = 15$ kV; a temperature of 5 ± 1 eV was obtained in this calculation.

During the time of the discharge, T_e remains practically constant within the limits of experimental error. A calculation of T_e for the discharge in the "Tuman" device [8] was made with the same method and gave a value of about 8 eV. This is in rather strong agreement with the value $T_e = 6 \pm 1$ eV which was derived from measurements of the plasma-string conductivity.

The correlation of the transmitted 8-mm UHF signal and the light emission from the plasma was considered along with the T_e determinations. To this end, simultaneous measurements of the scattered UHF signal and the light emission from the plasma were made in a certain chamber cross section. A strong correlation between the two quantities was observed up to frequencies of the order of several kilohertz.

Figure 1 shows an oscillogram of (1) a light signal and (2) the UHF signal used in the measurements. The observed signal correlation indicates that changes in the intensity of the plasma radiation result basically from changes in the plasma concentration rather than from changes in the plasma temperature.

The plasma inhomogeneities were investigated in the following fashion. Two light guides connected to a photomultiplier were aimed at the plasma. The two light guides were spaced 30, 60, 90, 120, 180, and 240 mm in the direction parallel to the chamber axis and in a meridional cross section, and 2, 5, 10, and 22 mm in the direction perpendicular to the chamber axis. The direction of the toroid axis was parallel to the direction of the external magnetic field H_z. The divergence of the collimated light beams incident on the photomultiplier was $\leq 5°$. The photomultiplier signals were applied to the S15-3 differential amplifier of an S1-17 oscilloscope. The amplifier channels were balanced with the aid of an accurate pulsed light source. The

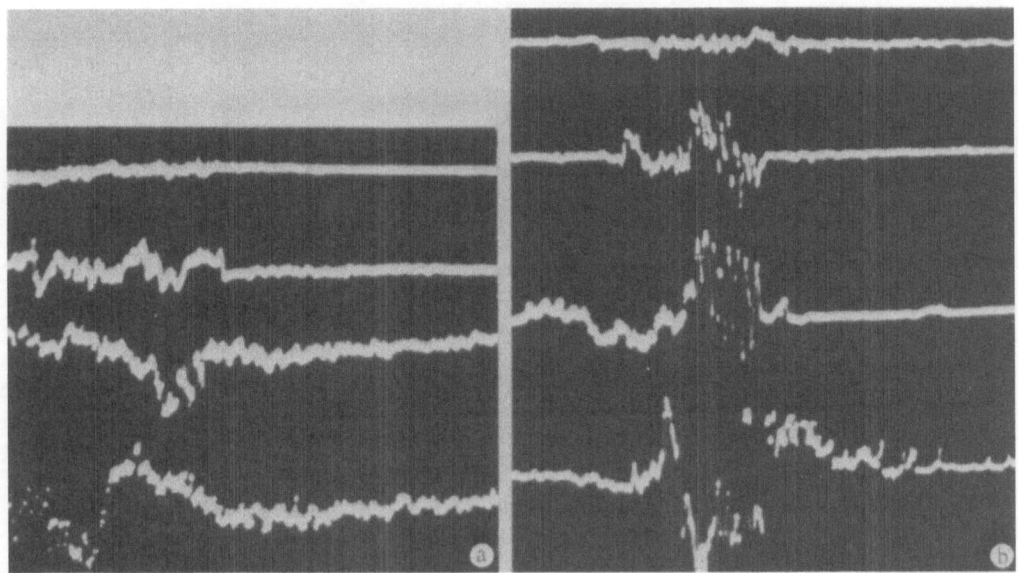

Figure 2. Oscillogram of the differential light signals derived at various points along the direction of the chamber axis; distances between the lightguides: 0, 60, 120, and 240 mm (a); as before, points in the direction perpendicular to the chamber axis; distances between the lightguides: 2, 5, 10, and 22 mm (b).

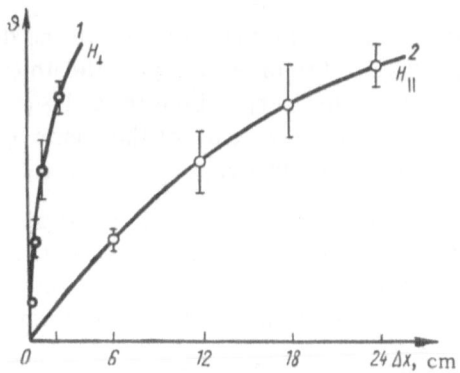

Figure 3. Average difference of the light signals as a function of the distance (2) in the direction parallel to the chamber axis; and (1) in the direction perpendicular to the chamber axis.

curve representing the signal difference during the establishment of the channel balance coincided with the null line.

It could be shown that the oscillations of the difference signal increase with increasing distance between the two light guides. Figure 2a shows oscillograms of the difference signals obtained when the distance between the light guides amounted to 0, 60, 120, and 240 mm in the direction parallel to the chamber axis; Figure 2b refers to the difference signals which were observed when the distances between the light guides amounted to 2, 5, 10, and 22 mm in the direction perpendicular to the chamber axis.

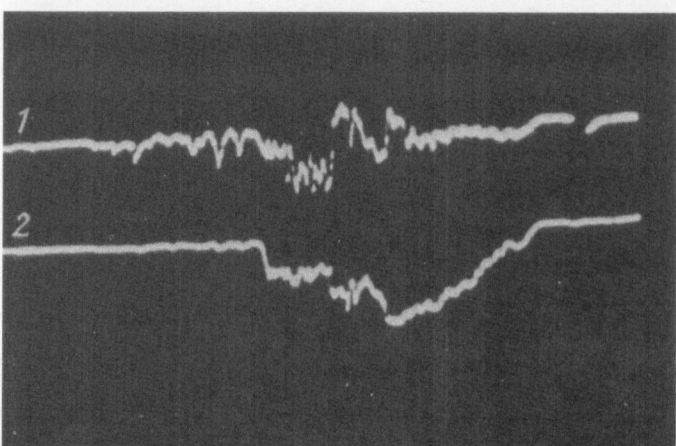

Figure 4. 1) Decrease in the difference signal in the
region of (2) the maximum of the total light emission.

The absolute values of the average deviation ΔI of the difference signal during the dis-

charge time, $\vartheta = \frac{\Sigma |\Delta I|}{n}$ were determined for each oscillogram.

The lower curve of Figure 3 shows the dependence of ϑ upon the distance in the direction parallel to the axis, whereas the upper curve refers to the distance in the direction perpendicular to the chamber axis.

Obviously, the inhomogeneities in the plasma increase in the direction perpendicular to the magnetic field at a higher rate than in the direction parallel to the magnetic field. The scale of the inhomogeneities in the direction perpendicular to the magnetic field is approximately one order of magnitude smaller than the scale of the inhomogeneities parallel to the magnetic field, as can be inferred from the curves.

When the transition from the low-current operating conditions to the high-current operating conditions takes place without a sharp increase in the current, the oscillograms representing the plasma emission are usually characterized by a single maximum. The oscillations of the difference signal decrease when the maximum intensity of the plasma emission is reached.

This effect seems to result from an increased homogeneity of the plasma at increasing plasma concentrations and is of definite interest for the purposes of plasma diagnostics.

REFERENCES

1.   A. B. Berezin et al., Transactions of the Seventh International Conference on Ionization Phenomena in Gases, Belgrad, 1965.
2.   S. E. Frish, Vestn. Leningr. Universiteta, No. 8, p. 129 (1953).
3.   V. E. Golant et al., Zh. Tekh. Fiz., 36:68 (1966).

MEASUREMENT OF THE CORRELATION FUNCTIONS
OF THE LIGHT-EMISSION INTENSITY
IN A PLASMA-BEAM DISCHARGE

E. V. Lifshits and E. A. Kornilov

As has been shown in both theory and experiment [1-3], spectroscopic techniques are very effective for investigations of processes occurring in a plasma-beam discharge and for determining the principal parameters of a plasma which is the result of the development of collective instabilities.

Of particular interest are investigations of the processes which occur in the transition from laminar to turbulent plasma states and from laminar plasma states to stationary turbulence in a plasma-beam discharge. The principal parameters of those processes can be derived from the correlation functions of the current density in the plasma and the electric field strength.

Direct observations of the electric fields or the density changes in the plasma as functions of time are a very effective method in turbulent plasma research. Experimental measurements of the correlation functions have been performed recently with the aid of techniques known from radio physics [4, 5]; apart from these, direct observations of the time dependence of the electric fields were made. Naturally, the question of using optical methods in those investigations arises. The main advantage of optical methods is that they do not introduce disturbances into the plasma. This is particularly important when one observes various types of radio-frequency oscillations which are excited when an instability develops, because some of these oscillations, particularly the various types of longitudinal waves, are not emitted from the plasma and one must introduce into the plasma probes, antennas, or other devices for the observation of these oscillations. Naturally, this implies strong disturbances of the conditions ruling in the plasma. Optical methods are completely free of these shortcomings. When photoelectric recording with a very small time constant is used, or when electron-optical converters are employed, one can investigate both the high-frequency and the low-frequency oscillations of complicated forms and with a strong time dependence. The oscillations generated in the transition from the laminar to the turbulent plasma state or from the laminar state to stationary turbulence are among those oscillations. In the corresponding experiments, one measures the correlation function of the radiation intensity and analyzes and observes directly the time-dependent changes of the light intensity.

The relation between the radiation intensity and the plasma parameters (density, temperature, and distribution function) is usually very complicated [1]. However, one can frequently single out a plasma parameter which determines the radiation intensity. Changes in the plasma density determine certain types of oscillations. The radiation intensity is, in turn, determined by the plasma density. In other cases, changes of the electron energy in the field of an exciting wave are important. Changes in the radiation intensity as a function of time result then from time-dependent changes of the average electron velocity. Obviously, when one measures simultaneously the correlation functions for white light and for a particular spectral line, one can determine the contribution of the various plasma parameters (density, velocity, etc.) to the changes in the light intensity.

The purpose of our work was to investigate the possibility of measuring correlation functions which relate the radiation intensity to low-frequency plasma oscillations in a plasma-beam discharge. Direct observations of the light intensity were made for this purpose and the time dependence of the light intensity was compared with the time dependence of oscillations measured with methods of radio physics.

The experiments were made with an electron beam (electron energy 5 keV and current 45 mA). The beam was injected into a glass bulb 10 cm in diameter and 40 cm in length; this bulb was placed into a magnetic field having a strength of up to 2000 G. The setup was similar to that described in [6]. By changing both the magnetic field strength and the pressure in the bulb, one can obtain conditions for which the high-frequency fields (which are excited when a beam instability develops) create and maintain a plasma density which is 3-4 orders of magnitude greater than the density of the injected beam ($n_p \sim 10^{11}$-10^{12} cm^{-3} and $n_b \sim 10^8$ cm^{-3}).

Under these conditions, intensive low-frequency (10-200 kHz, depending upon the gas) oscillations can be observed along with high-frequency oscillations (600-6000 MHz). The low-frequency oscillations strongly affect the processes in the plasma-beam discharge [6]. The low-frequency oscillations can be observed at the location of sensors placed into the plasma, as well as in the light emission from the plasma.

The main goal of our work was to establish the relation between the oscillations of the light emission and the oscillations appearing in the sensor current. Therefore, the signals derived from the sensors (photomultiplier, which recorded the intensity of the total light flux or of a single spectral line singled out with a UM-2 monochromator) were applied to an S1-16 oscilloscope. An FEU-64 photomultiplier was used for recording light fluxes.

The correlation between the signals under investigation can be estimated when one applies one of the signals to the synchronization stage of the oscilloscope. When a full correlation between the signals exists, a stationary waveform is displayed on the screen. When no correlation exists, synchronous triggering can be used to obtain an idea of the overall form of the oscillations and their phase relations.

In order to determine accurately the signal correlation, the time-dependent autocorrelation functions X(t) of the oscillations in the current of a sensor placed into the plasma and Y(t) of the oscillations in the light emission were considered. We recall that these autocorrelation functions are defined by the relations

$$R_{xx}(\tau) = \frac{1}{T} \int\limits_{\lim T \to \infty}^{T} \!\!\!\!\!\!\!\! {}_0 \;\; X(t) X(t+\tau)\, d\tau,$$

$$R_{yy}(\tau) = \frac{1}{T} \int\limits_{\lim T \to \infty}^{T} \!\!\!\!\!\!\!\! {}_0 \;\; Y(t) Y(t+\tau)\, d\tau.$$

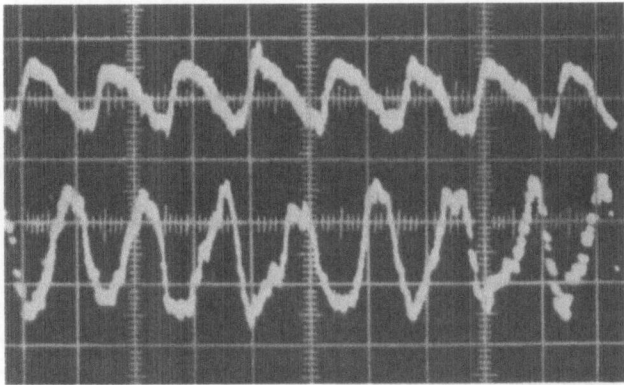

Figure 1. **Oscillations in the sensor current** (upper oscillogram) and in the intensity of the light emission from the plasma (lower oscillogram) for a small oscillation amplitude. Field strength, 2-3 V/cm; pressure $p \simeq 8 \cdot 10^{-4}$ mm Hg; $f \approx 10$ kHz; time base, 10 μsec/cm.

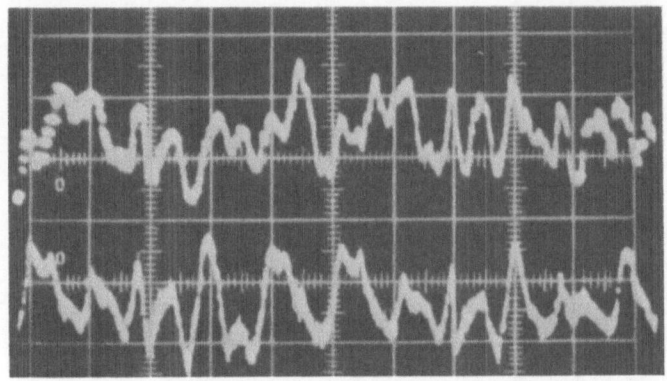

Figure 2. Oscillations in the sensor current (upper oscillogram) and in the intensity of the light emission from the plasma (lower oscillogram) for a great oscillation amplitude. Field strength, ~15-20 V/cm; $p \simeq 2 \cdot 10^{-4}$ mm Hg; time base, 10 μsec/cm.

The autocorrelation functions can be used to determine the form of the oscillations, their correlation time, and their spectral density [7].

The time-dependent autocorrelation functions were recorded with an IDK-3 correlator which enabled us to observe signals in the frequency interval 0-50 kHz.

Earlier investigations [6] have shown that the amplitude of all oscillations in a plasma-beam discharge depends nonmonotonically upon the pressure. The amplitude increases suddenly at $P > P_{cr}$ and decreases afterwards with increasing pressure. Thus, by changing the pressure, one can observe the form of the oscillations at various signal levels and degrees of plasma turbulence.

With the beam parameters used in our experiments (U = 5 keV and I_b = 45 mA), the maximum amplitude of the low-frequency oscillations in argon was observed at $7 \cdot 10^{-5}$ mm Hg (at a field strength of about 20 V/cm).

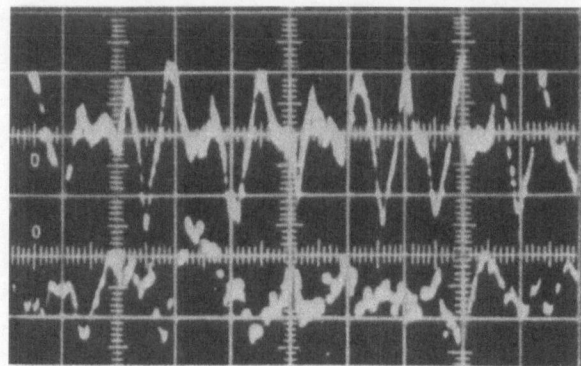

Figure 3. Oscillations in the sensor current (upper oscillogram) and in the intensity of the Ar II 4806 Å line emitted from the plasma (lower oscillogram) for a great oscillation amplitude. Field strength, ~10–15 V/cm: $p \approx 4 \cdot 10^{-4}$ mm Hg; time base, 10 μsec/cm.

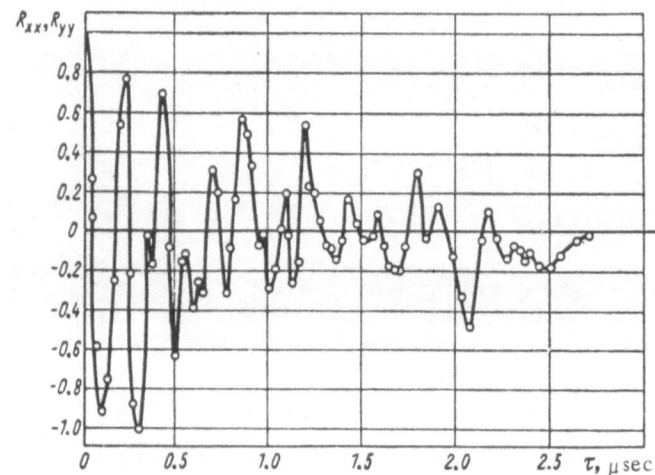

Figure 4. Time-dependent autocorrelation functions of the oscillations in the sensor current and the oscillations in the light emission from the plasma. Pressure $p \approx 6 \cdot 10^{-4}$ mm Hg.

It follows from the oscillograms of the oscillations that the oscillations change from almost sinusoidal oscillations to noise peaks at increasing oscillation amplitudes. A full correlation is observed between the oscillations in the current of sensors inside the plasma and the oscillations in the light emission. The two oscillations differ insofar as the phase of the light-flux oscillations is shifted by 180° relative to the oscillations in the detector current (Figures 1 and 2). The time dependence of the intensity of individual spectral lines is completely different. At small oscillation amplitudes, a full correlation similar to that shown in Figure 1 exists between the oscillations in the sensor current and the oscillations in the intensity of light emission. No correlation can be observed at great amplitudes.

Figure 3 shows oscillograms of the oscillations picked up with the sensor (upper oscillogram) and of oscillations in the intensity of the Ar II 4806.07 Å line (lower oscillogram). The amplitudes of the oscillations shown in Figures 2 and 3 are almost identical. It follows

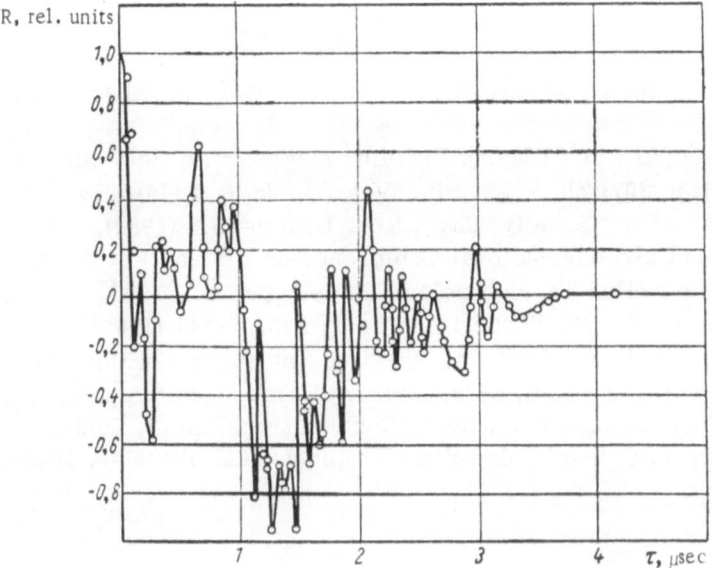

Figure 5. Time-dependent autocorrelation function of
the oscillations in the intensity of the Ar II 4806 Å
line emitted from the plasma. Pressure $p \simeq 6 \cdot 10^{-4}$
mm Hg.

from a comparison of the oscillograms shown in Figures 2 and 3 that the total light-flux oscillations and the sensor-current oscillations agree completely in both form and frequency,
whereas components with higher harmonics are observed in individual spectral lines.

A similar pattern is found when one compares the time-dependent autocorrelation functions
of the oscillations in the sensor current, the total light flux, and the individual spectral lines.
The autocorrelation functions of the oscillations in the total light flux and the oscillations in the
sensor current are completely identical (Figure 4), whereas the autocorrelation functions for individual spectral lines are different (Figure 5).

Inspection of the time-dependent autocorrelation functions of oscillations in the total
light flux and in the Ar II 4806.07 Å line indicates that the correlation time is about 9.8 μsec in
the first case and 3.5 μsec in the second. Harmonics with higher frequencies than those of
the autocorrelation function of the total light flux are found in the autocorrelation function of
the single spectral line.

Thus, the time-dependent changes of the intensity of the total emission and of the electric fields in the plasma are completely correlated. This leads to the conclusion that the low-
frequency oscillations in a plasma-beam discharge can be investigated by measuring the total
light emission from the plasma.

The differences in the time dependence of the total light-flux intensity and the intensity
of individual spectral lines result most likely from the fact that the intensity of a single spectral
line is more dependent upon changes in the energy distribution of the plasma electrons than the
intensity of the total light flux [8]. One can therefore assume that simultaneous measurements
of the autocorrelation functions of the light flux intensity and the intensity of individual spectral lines allow determinations of certain properties of the electron distribution function of
the plasma. This problem will be the subject of our future investigations.

In conclusion, we express our gratitude to Ya. B. Fainberg and L. I. Bolotin for valuable
discussions and to A. G. Shevlyakov for his assistance in the measurements.

REFERENCES

1.   E. V. Lifshits, Optika i Spektroskopiya, 19:19 (1965); Zh. Eksper. i Teor. Fiz., 53:943 (1967).

2.   E. V. Lifshits et al., in: "Plasma Physics: Magnetic Bottles" [in Russian], Kiev, Naukova Dumka (1965), p. 207; Zh. Tekh. Fiz., Vol. 36, No. 6 (1966).

3.   L. D. Smullin and W. D. Getty, Phys. Rev. Letters, 9:1 (1962).

4.   A. K. Berezin et al., International Conference on Plasma Physics and Investigations in the Field of Controlled Thermonuclear Fusion, Culham, 1965.

5.   E. A. Kornilov et al., Zh. Eksperim. i Teor. Fiz., 4:147 (1966).

6.   E. A. Kornilov et al., in: "Plasma Physics: Interaction of Charged Particle Beams with a Plasma" [in Russian], Kiev, Naukova Dumka (1965), pp. 24, 36.

7.   F. Lange, "Correlation Electronics" [Russian translation], Leningrad, Sudpromgiz (1963).

8.   S. M. Hamberger, Culham Laboratory Preprint CLM-P8 (May, 1962).

APPLICATION OF A CORRELATION TECHNIQUE
FOR INVESTIGATIONS OF THE TURBULENT PLASMA
IN THE ALPHA DEVICE

V. A. Rodichkin, G. A. Serebrenyi, and A. M. Timonin

A large number of collective oscillations which result from simultaneous, ordered motions of all particles in a system or of particle subsystems [1-3] are one of the most general properties of a turbulent plasma. The plasma parameters resulting from the superposition of those oscillations, e.g., the electric and magnetic fields, the plasma density, the temperature, the spectral characteristics, etc., are random functions of time. Under these conditions, multidimensional probability-distribution functions of these quantities can provide full information on a turbulent plasma. The measurement of these functions encounters great experimental and computational difficulties. Thus, in the majority of cases, fluctuations are disregarded and only average values of the plasma parameters are measured. Knowledge about the average values alone can provide only a rough idea of the processes occurring in the plasma because the fluctuations include information on collective plasma oscillations.

A better idea of these processes, particularly those observed in research on plasma instabilities and the resulting collective oscillations, can be obtained from measurements of the correlation functions of the fluctuating quantities [4, 5]. In the general case, a correlation function is a product averaged over events:

$$\varphi(\tau) = \overline{x_1(t)\, x_2(t + \tau)},\tag{1}$$

where $x_1(t)$ and $x_2(t)$ denote the center values of the random processes $X_1(t)$ and $X_2(t)$ and the bar refers to averaging over the events.

Correlation functions are easily derived for stationary stochastic processes for which one can use the ergodic theorem and replace the averaging over events with an averaging over time for a single event so that the expression for the correlation function assumes the form

$$\varphi(\tau) = \lim_{T \to \infty} \frac{1}{2T} \int_{-\infty}^{+\infty} x_1(t)\, x_2(t + \tau)\, dt.\tag{2}$$

However, in the majority of cases, the processes in a plasma are nonstationary. Thus, the ergodic theorem cannot be applied. However, when a stochastic process is quasi-stationary, then, as before, one can use Eq. (2) and introduce a short-time correlation function $\varphi(t, \tau)$, which depends only upon the time. A quasi-stationary process is defined as a process for which

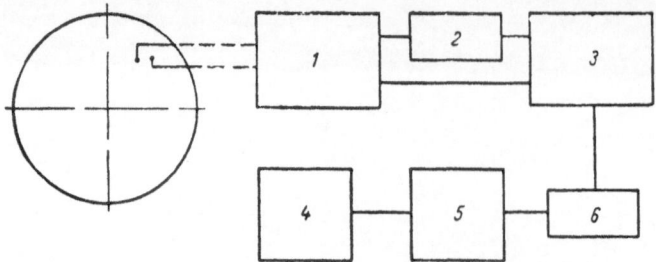

Figure 1. Block diagram for correlation measurements. 1) Input transformers and attenuators; 2)
delay line; 3) multiplying circuit; 4) oscilloscope; 5)
amplifier; 6) integrating circuit.

the characteristic time of changes in the function x(t) is much smaller than the time during which
the mean square value of that function changes.

When one knows the correlation function, one can rather easily determine the power spectrum of a fluctuation process. According to the Wiener−Khinchin theorem, the correlation
function and the power spectrum are related by the Fourier transform:

$$S(f) = \int_{-\infty}^{+\infty} \varphi(\tau) e^{-j\omega\tau} d\tau. \tag{3}$$

In addition to the power spectrum, one can determine with the correlation function the relation between fluctuations of various plasma parameters, the size of plasma inhomogeneities,
the velocity of plasma motions, and other effects which will be described below.

It is generally accepted [6] that a discharge in a weak longitudinal magnetic field (particularly the discharge type of the Alpha device [7]) is characterized by an extensive spectrum
of instabilities and oscillations. However, the measurements of [8] and [9] have shown that
low-frequency plasma oscillations in the frequency range reaching to 1 MHz are most clearly
visible. Apart from this, low-frequency oscillations are very safe as far as a plasma con-
tainment is concerned. It is therefore of primary interest to investigate the spectrum of the
plasma turbulence in this particular frequency range. To do this, a correlation instrument
whose block diagram is shown in Figure 1 was assembled. The receiver was adapted to the
Alpha device in order to investigate fluctuations of both the electric and magnetic fields which
were measured with electric and magnetic sensors. A ring modulator with D1B diodes was
used as the multiplying element of the correlator. The integrating circuit comprised a simple
RC circuit with a time constant of about 0.5 msec. An RC filter in the form of a "three-pole"
[10] was inserted at the output of the correlator. This RC filter and the integrating RC circuit formed a low-frequency filter with an upper frequency limit of about 300 Hz. This frequency agrees with the principal frequency of the discharge pulse in the Alpha device. Standard LZT-4-1200 delay lines having a maximum delay time of 20 μsec and a minimum delay
time of 0.2 μsec (disregarding the delay "zero") were used in the receiver. It is therefore
possible to investigate periodic processes in the frequency interval ranging from 15 kHz to
1 MHz (resolving power of the correlator). The total frequency band transmitted by the correlator ranges from 5 kHz to 1 MHz. The signal derived from the receiver was displayed on
an oscilloscope and recorded on photographic paper. Typical oscillograms of the signal obtained with the correlator for various τ are shown in Figure 2.

As has been mentioned above, the autocorrelation function makes it possible to establish
the energy spectrum of a fluctuation process. Figure 3a shows the normalized autocorrelation

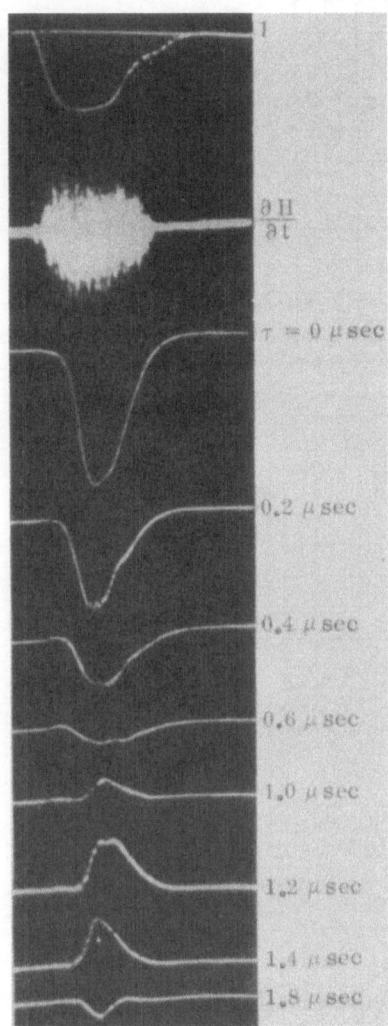

Figure 2. Correlation curves of the derivative of the magnetic field $\varphi_{H'H'}(t, \tau)$ for various τ values.

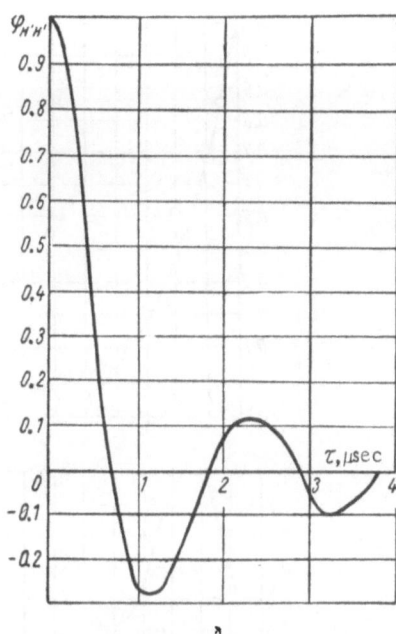

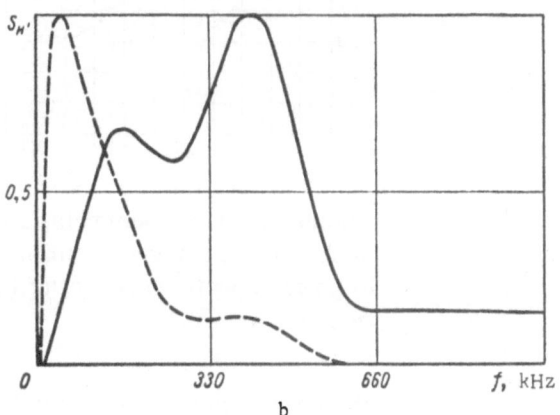

Figure 3. a) Autocorrelation function $\varphi_{H'H'}(t_0, \tau)$ of the signal derived from the magnetic sensor; b) power spectrum $S_{H'}(f)$ of the derivative of the magnetic field fluctuations; the dashed line indicates the power spectrum $S_H(f)$ of the magnetic field fluctuations.

function of the derivative of the transverse magnetic field component, whereas Figure 3b displays the corresponding power spectrum. In addition, Figure 3b shows the power spectrum of the magnetic field calculated from the spectrum of the derivative (dashed line). Figure 4 shows the correlation function of the signal derived from an electric double sensor and the corresponding power spectrum. The measurements were made in the center of the current string at the time t_0 at which the current in the plasma reached its maximum value. The spectra shown in the figures were obtained on a computer when a cosine-Fourier transformation of the autocorrelation functions was used. The frequency range in which the oscillation energy is concentrated, is clearly visible on the power spectra.

Apart from the autocorrelation function, the correlator can be used to measure the correlation of fluctuations of various plasma parameters.

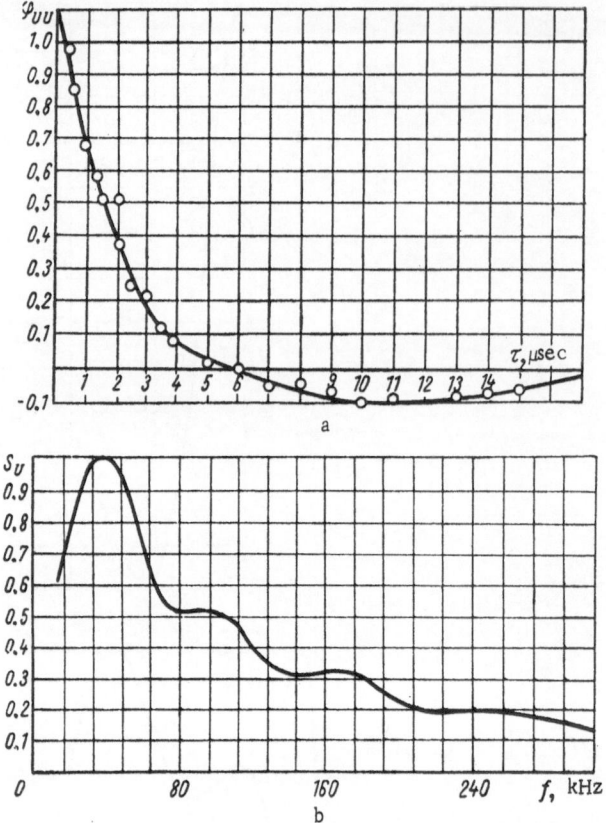

Figure 4. a) Autocorrelation function $\varphi_{UU}(t_0, \tau)$ of the signal derived from the two electric sensors; b) power spectrum $S_U(f)$ of an electric sensor.

The concept of the correlation function can be used to determine the size of spatial plasma inhomogeneities. Figure 5 shows the relation between the correlation coefficient of the signals derived from two magnetic coils and from two electric sensors as a function of the distance between these sensing elements (distance in the direction of the chamber radius). The figure indicates that the correlation coefficient decreases with increasing distance. When one determines the size of an inhomogeneity from the half-width of the correlation function, one obtains a size of 4-5 cm in the high-frequency region (200-400 kHz) which is represented by the autocorrelation function of the magnetic field derivative. In the range of lower frequencies (40-80 kHz), which can be represented by the autocorrelation functions of the magnetic field and the signal derived from the two sensors, one observes inhomogeneities with a size of 7-8 cm. When the plasma inhomogeneities move with an average velocity v, this results in a time shift between the two signals derived from the sensors spaced by the distance l. In this case, one can set in Eq. (1)

$$x_2(t + \tau) = \alpha x_1\left(t \pm \frac{l}{v} + \tau\right) + \delta x_2(t + \tau), \qquad (4)$$

where α denotes some proportionality coefficient, and $\delta x_2(t)$ is the part of the signal $x_2(t)$ which is uncorrelated with $x_1(t)$. Then, we have

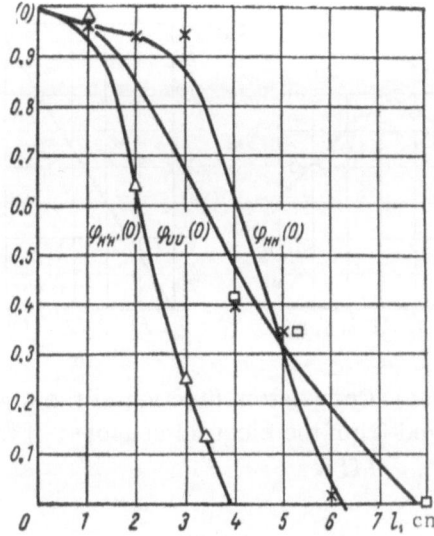

Figure 5. Relation between the correlation coefficients and the distance between the sensors.

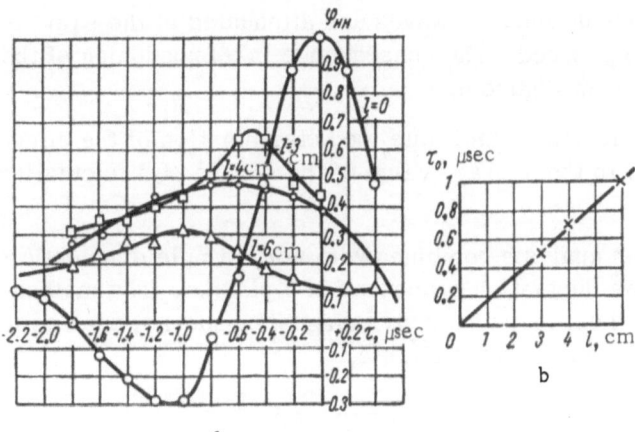

a

b

Figure 6. a) Correlation functions $\varphi_{H_1'H_2'}(\tau)$ of the signals derived from the two magnetic sensors; b) the relation $\tau_0 = f(l)$.

$$\varphi(\tau) = ax_1(t)\,x_1\left(t\pm\frac{l}{v}+\tau\right) + x_1(t)\,\delta x_2(t+\tau) = ax_1(t)\,x_1\left(t\pm\frac{l}{v}+\tau\right), \tag{5}$$

because the noncorrelated part δx_2 of the signal does not contribute to the correlation function. The right side of Eq. (5) contains a quantity which is proportional to the autocorrelation function of the random process $x_1(t)$. This function assumes its maximum value for

$$\pm\frac{l}{v} + \tau_0 = 0, \tag{6}$$

because $\varphi(\tau) \leq \varphi(0)$. The sign of τ_0 for which condition (6) is satisfied is determined by the direction of motion of the inhomogeneities. After determining the τ_0 value for which $\varphi = \varphi_{max}$, one can determine $v = l/\tau_0$. Figures 6 and 7 show the correlation functions of the

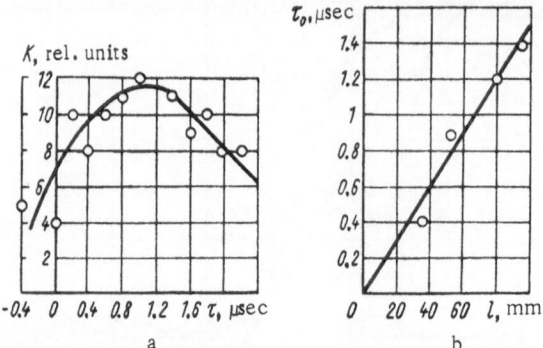

Figure 7. a) Correlation function of the signals derived from the electric sensors; b) the relation $\tau_0 = f(l)$.

signals derived from the magnetic and electric sensors when the distances between the detectors were varied. In addition, these figures show the relation $\tau_0 = f(l)$. It follows from the curves that the velocity of motion of the inhomogeneities amounts to $(5-6) \cdot 10^6$ cm/sec. These conclusions are applicable when l is smaller than the attenuation length of the waves. Since the disturbances are attenuated in the course of time, an increase in l is accompanied by a decrease in the correlation, and since the waves are attenuated at the same time, mainly the high-frequency components are reduced. The consequence is a broadening of the correlation functions, as can be inferred from Figure 6.

The autocorrelation functions facilitate the determination of the effective fluctuation values, because $\varphi(0)$ is equal to the average value of the square of a fluctuating quantity or to the dispersion.

Thus, the correlation analysis combined with very simple diagnostic means provides exhaustive information on the fluctuation processes in a plasma. The method can also be employed in the analysis of other, more extensive diagnostic means, such as optical methods, methods involving microwaves or particles, etc.

REFERENCES

1.   B. B. Kadomtsev, in: "Problems of Plasma Theory" [in Russian], M. A. Leontovich, ed., Moscow, Atomizdat (1964), Vol. 4, p. 188.
2.   A. A. Vedenov, "Theory of Turbulent Plasmas" [in Russian], Moscow, VINITI (1965).
3.   A. I. Akhiezer et al., "Collective Plasma Oscillations" [in Russian], Moscow, Atomizdat (1964).
4.   B. R. Levin, "Theory of Stochastic Processes and Application of the Theory in Radio" [in Russian], Moscow, Sovetskoe Radio (1957).
5.   F. Lange, "Correlation Electronics" [Russian translation], Leningrad, Sudpromgiz (1963).
6.   B. B. Kadomtsev, Nuclear Fusion, Suppl., Part 3, p. 969 (1962).
7.   V. A. Glukhikh et al., Zh. Tekh. Fiz., 30:1394 (1960).
8.   M. J. Rusbridge et al., Nucl. Fusion, Suppl., Part 3, p. 895 (1962).
9.   A. N. Saidel' et al., Zh. Tekh. Fiz., 30:1437 (1960).
10.   E. O. Saakov, "Theory and Calculation of Selective RS Systems" [in Russian], Moscow, Gosenergoizdat (1954).

INVESTIGATION OF IONIC CYCLOTRON OSCILLATIONS IN A PLASMA BY MEANS OF A FAST-RESPONSE ANALYZER OF THE FREQUENCY SPECTRUM

A. V. Bortnikov, N. N. Brevnov,
V. G. Zhukovskii, and M. K. Romanovskii

When a plasma is created in a magnetic bottle by injection of fast particles, the velocity—space distribution of the trapped ions is highly anisotropic [1]. Longitudinal oscillations at the ionic cyclotron frequency and its harmonics can occur in these plasma anisotropies. The oscillations propagate almost perpendicular to the magnetic field lines, i.e., a cyclotron instability can develop [2, 3].

A pulsed analyzer was used with the AS device [5] to study the effect of an anisotropic particle distribution in velocity space upon the generation and rate of development of cyclotron instabilities.

Setup and Apparatus

Figure 1 shows schematically the AS device [4]. Direct current flowing through coils 4, 5, 6, 9, and 10 creates a homogeneous, stationary magnetic field with a "plug" in the area of coil 11. The magnetic field strength is $H_0 = 2000$ Oe. The "plugging" ratio for the stationary field is 1.5. A second magnetic "plug" is produced with a current pulse applied to coil 15. The distance between the magnetic plugs is $L = 200$ cm, and the chamber diameter 20 cm. The plasma was generated either by injecting a beam of atomic hydrogen ions with an energy of 10 keV through a magnetic channel or by injecting atomic ions with an energy of 2.5 keV from a source situated on the axis inside the chamber. In the first case, the fast ions formed a hollow cylinder (Figure 2a); in the second case (Figure 2b), the ions filled completely the cross section of the magnetic bottle. The ions were trapped in the magnetic bottle by increasing the magnetic field in coil 15 (see Figure 1). The injection current was as high as 10 mA, and the plasma density reached 10^8 ions/cm^3.

Plate and loop antennas were used to study the electromagnetic field of the oscillations. The signals derived from the antennas were applied to the input of a pulse analyzer. The principle of successive analysis with transformation of the spectrum was used in the analyzer. The analyzer described in [6] was the basis of the instrument. Figure 2 shows the block diagram of the analyzer.

The signal is passed through a matched cable and an attenuator and arrives at the input of a broad-band preamplifier. The signal is then fed from the preamplifier to a superhetero-

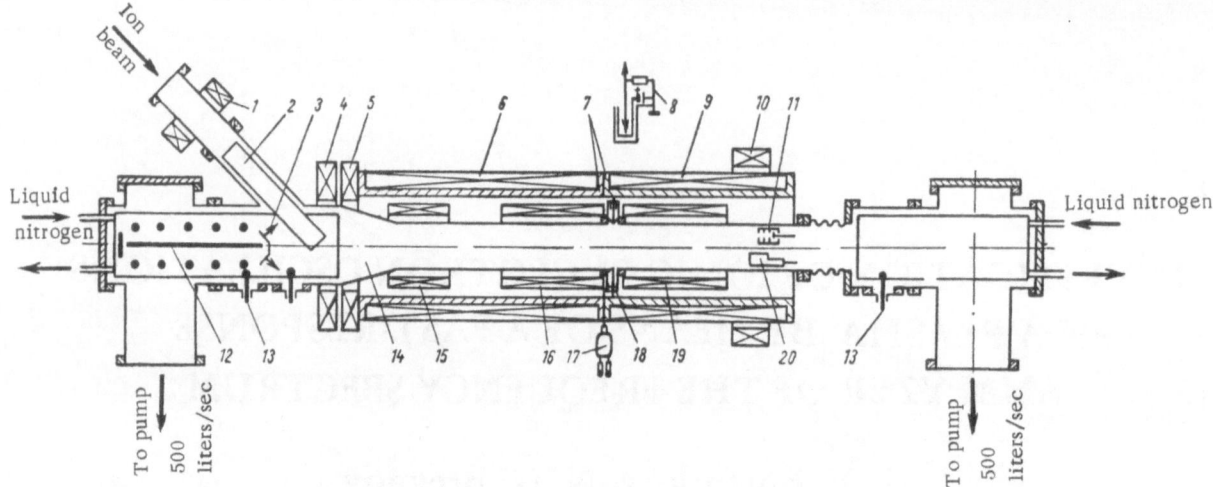

Figure 1. Scheme of the AS device. 1) Magnetic lens which focuses the ion beam; 2) channel;
3) nitrite; 4, 5, 6, 9, 10) coils for the stationary magnetic field; 7) copper shield; 8) neutral
particle detector; 11) analyzer for the energy spectrum of the secondary ions; 12) axial ion
source; 13) titanium evaporators; 14) chamber; 15) shutoff coil; 16, 19) compression coils;
17) palladium leak; 18) rodlike sensor; 20) current pickup.

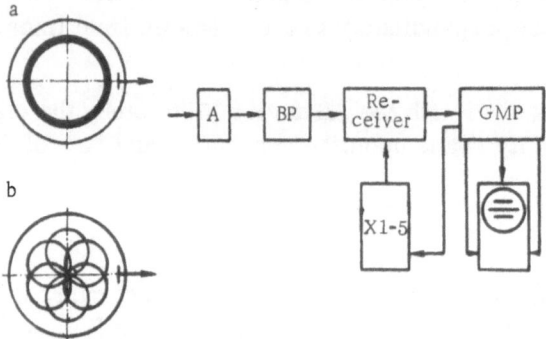

Figure 2. Block diagram of the instrument.
a) Form of the plasma formation when ions
are injected through the magnetic channel; b)
form of the plasma when ions are injected
from a source within the chamber; A) attenuator; BP) broad-band preamplifier; GMP) generator of modulating pulses.

dyne receiver. In order to perform an analysis in a very short time, the transmission band
of the receiver has been increased to 600 kHz, and a frequency-modulated Kh1-5 generator is
used as the heterodyne stage.

In order to modulate the Kh1-5 generator and to display the spectrum on the oscilloscope
screen as a function of time, a generator delivering several series of different pulses was
developed. The generator comprises a sawtooth generator and a step pulse generator. Figure 3 shows the generator circuit. The sawtooth consists of tubes 2, 3, and 4 and can be operated continuously or in standby operation. The number of sawtooth peaks in standby operation
(1 to 10 peaks) is given by the duration of the pulses which are derived from the right side of

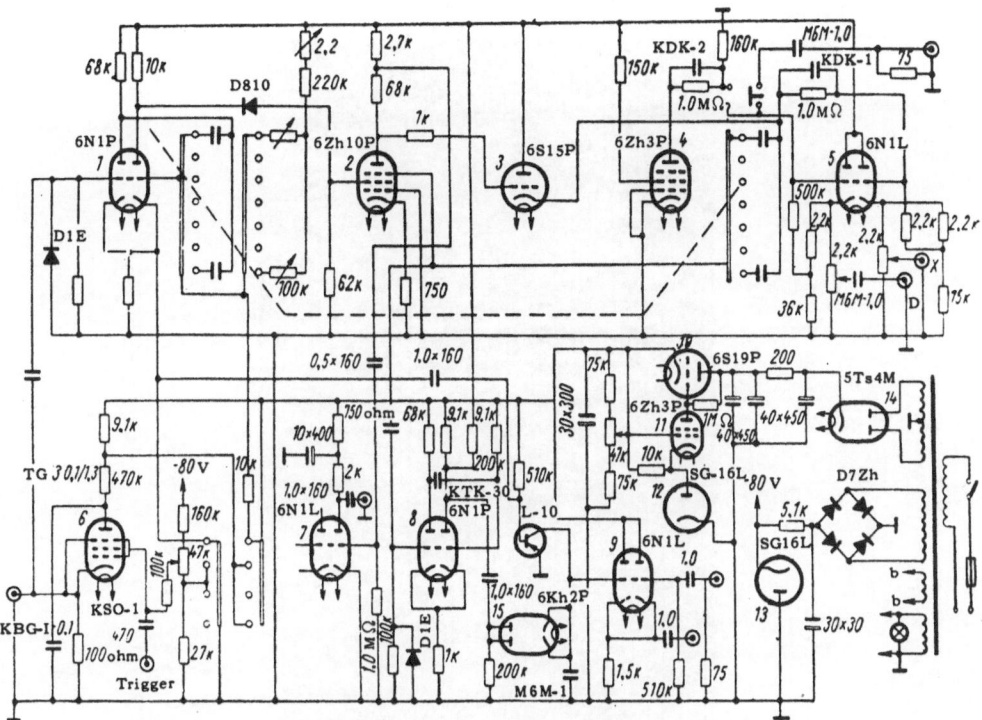

Figure 3. Circuit for the generator of modulating pulses.

standby multivibrator 1. The signal derived from the sawtooth generator via a matched cathode follower 5 is applied to the time base of an ENO-1 oscilloscope (X output) and is used to modulate the Kh1-5 generator (output D). A pulse-shaping stage 7 provides the blanking pulses for the oscilloscope which correspond to the ramp sections of the sawtooth waveform.

The pulses corresponding to the return of the sawtooth waveform are derived from the cathode of tube 2 and applied to a multivibrator 8 which causes a step-like charging of a capacitor. In order to obtain several spectra on the oscilloscope screen, the signal under inspection is shifted by means of a step pulse in cathode follower 9 and applied to the input of an ENO-1 oscilloscope. An internal trigger generator 7 is provided in the circuit.

With a receiver bandwidth of about 0.6 MHz one can perform an analysis in the frequency range 1 to 18 MHz within a time of about 100 μsec. Up to ten spectra can be displayed simultaneously (Figure 4). The analyzer has a sensitivity of 5 μV.

In order to investigate the oscillation amplitude as a function of time at a fixed (ionic cyclotron) frequency, the receiver signal is applied to an S1-17 oscilloscope, and a GSS-6 generator is used as a heterodyne stage.

Results of the Measurements

Figure 5 shows an oscillogram of the oscillation spectrum obtained with an electrostatic antenna in the center of the device and with a plasma in the form of a hollow cylinder. Three harmonics of the ionic cyclotron frequency are clearly visible on the spectrum. The amplitude of the harmonics changes as a function of time.

Figure 6b is an oscillogram which shows the development of the oscillations with the cyclotron frequency. When fast ions are trapped in the bottle by a pulsed magnetic plug, the oscillation amplitude increases at the beginning (in the time increment of 20-30 μsec), and

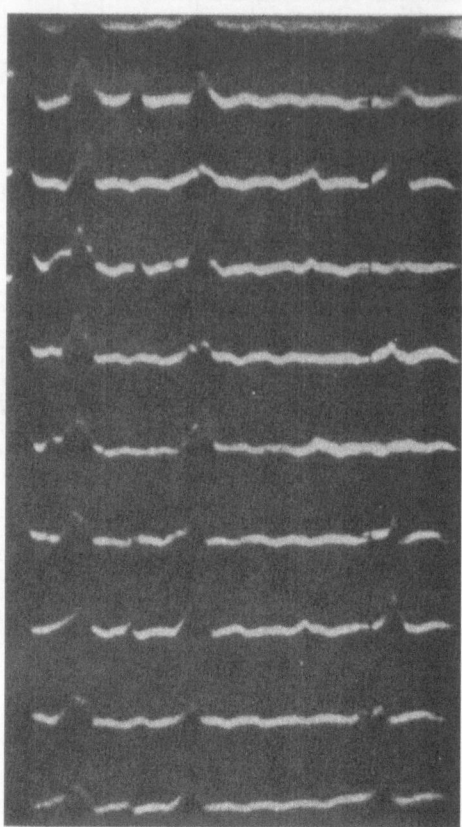

Figure 4. Spectrum consisting of ten
traces with 3, 8, 10, and 20 MHz marks.
Duration of the time base of each trace,
100 μsec.

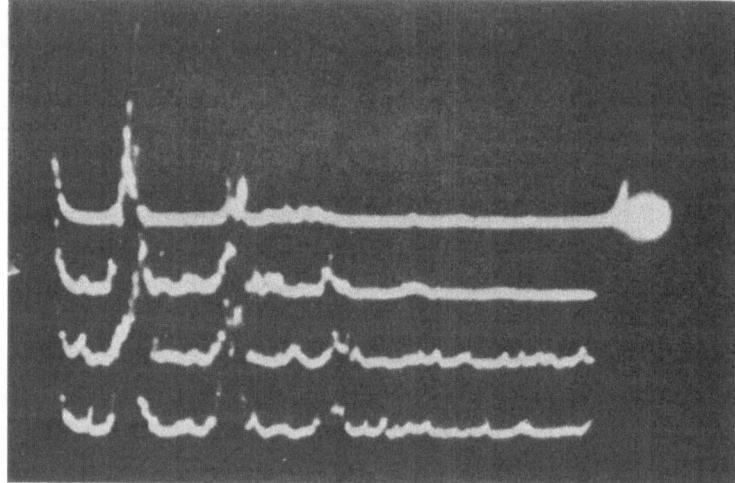

Figure 5. Oscillation spectrum obtained with an electrostatic antenna mounted in the chamber center. Harmonics
at 3.8, 7.6, and 11.4 MHz.

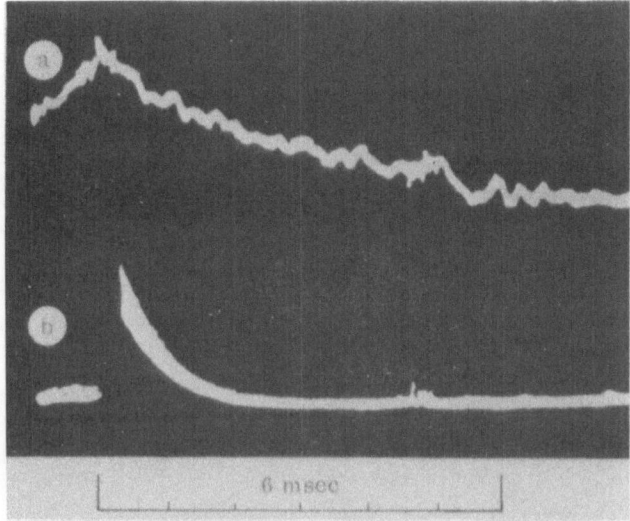

6 msec

Figure 6. Oscillograms of the signals obtained (a) from the neutral particle detector when ions were injected through the magnetic channel and (b) from the electrostatic antenna mounted in the chamber center after shutting off the plasma from the injector (principal harmonic).

then the attenuation of the oscillations sets in. The signal appearing at the neutral particle detector (Figure 6a) decreases at a rate three times slower. This ratio of the signal attenuation time and the densities remains conserved when the pressure changes from 10^{-8} to $3 \cdot 10^{-7}$ torr [5].

Measurements of the ion lifetime revealed that the lifetime amounts to 3-5 msec and depends only upon the charge exchange of the fast ions with neutral gas atoms. Thus, the development of the oscillations does not involve the exit of a considerable number of ions from the bottle.

The amplitude of the cyclotron oscillations before and after shutting off the plasma injection is proportional to the density of the particles trapped in the device. This indicates that the fields generated are coherent. The electric field strength of the oscillations amounts to about 1 V/cm at maximum.

Figure 7 (curve 1) shows the dependence of the density of the trapped, fast ions upon the stationary magnetic plug parameter. The current of fast ions incident on the current pickup behind the stationary plug was measured while the plugging ratio was varied. After that, a correction was introduced which takes into account the time needed by the ions to travel through the magnetic plug to the exit. By differentiating the curve, the density distribution function of the particles (curve 2) can be obtained as a function of the plugging ratio. This function is similar to the longitudinal-velocity distribution function of the ions.

When the stationary plug is made less effective, the number of particles trapped in the device decreases gradually. Nevertheless, the amplitude of the oscillations excited in the electrostatic antenna remains constant to the plugging ratio R = 1.1, but decreases sharply thereafter (curve 3). This seems to indicate that in this case, the oscillations involve particles whose velocity vector is almost perpendicular to the magnetic field ($\alpha \sim 80°$).

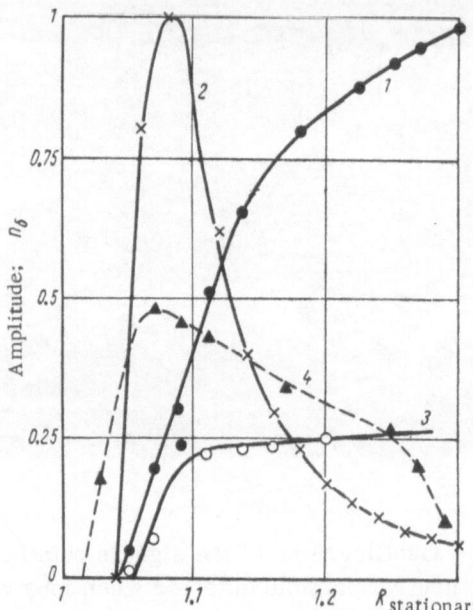

Figure 7. Relation between the plugging ratio
of the stationary plug and the following quantities: (1) density n_σ of the fast particles
trapped in the device when ions are injected
through the magnetic channel; (2) distribution
function dn_σ/dR; (3) amplitude of the electric
field at the cyclotron frequency after shutting
off the plasma supplied by the injector; and (4)
distribution function in the case of ion injection from a source situated on the axis of the
device.

The oscillation spectrum obtained in the second mode of filling the chamber with a
plasma (Figure 2b) has the same form as the spectrum shown in Figure 5, before the plasma
is shut off from the source. The field strength of the oscillations per milliampere of injection current amounts to about 1 V/cm and is proportional to the density.

Though the signal derived from the antenna decreases rapidly when the plasma is shut
off from the source (Figure 8b), the signal persists for a longer time period at the neutral
particle detector (Figure 8a). The longitudinal-velocity distribution function of the particle
is in this case much broader (see Figure 7).

Estimates of the anisotropy parameter $\tau = T_{\perp i}/T_{\parallel i}$ [2] for this distribution function
result in a value which is smaller than that for the ion distribution function (see Figure 7,
curve 2). This result can be explained by the fact that no oscillations with the cyclotron frequency develop after shutting off the plasma from the source, provided that the plasma fills
uniformly the cross section of the device.

The investigations which intend to clarify the influence of the various distribution functions of the injected ions upon the generation and development of cyclotron instabilities are
continued at the present time.

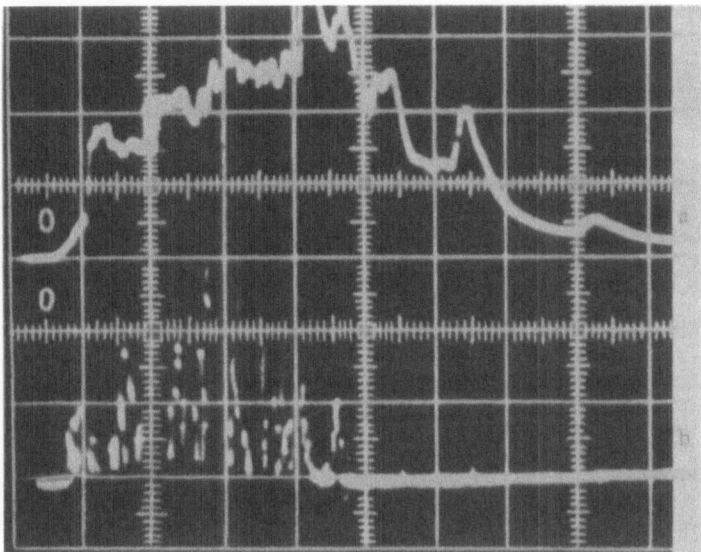

Figure 8. Oscillograms of the signals obtained upon ion injection from a source inside the device: (a) signal from the neutral particle detector and (b) signal from an electrostatic antenna located at the center of the chamber; first harmonic of the cyclotron frequency observed when the plasma was shut off from the source.

Conclusions

For investigating the time dependence of the oscillation spectrum, a pulsed analyzer was developed for the frequency range 1-18 MHz; the analyzer has an analyzing time of 100 μsec and allows the simultaneous observation of ten spectra.

The analyzer was used to investigate cyclotron oscillations for two modes of filling the AS device with a plasma. It could be shown that: (1) when the plasma has the form of a hollow cylinder, a cyclotron instability with an increment of 20-30 μsec develops; this instability stabilizes afterwards; the lifetime of the plasma is not affected by the development of the cyclotron instability; and (2) when the plasma uniformly fills the entire cross section of the device, the amplitude of the oscillations decreases rapidly once the plasma has been trapped.

In conclusion, we consider it to be our duty to express our gratitude to M. Pavshuk for developing the generator circuit, and to the service personnel for assistance in our work.

REFERENCES

1. G. F. Bogdanov et al., Nuclear Fusion, Suppl. (1962), Part 1, p. 215.
2. V. I. Pistunovich, Atomnaya Energiya, 14:72 (1963).
3. E. Harris, Phys. Rev. Letters, Vol. 2, No. 2 (1959).
4. A. V. Bortnikov et al., Atomnaya Energiya, 18:256 (1965).
5. A. V. Bortnikov et al., Preprint 988 of the Institute of Atomic Energy (1965).
6. A. N. Kharkov, Prib. i Tekhn. Eksperim., No. 5, p. 115 (1961).